THE STRATEGY TRAP

INDIA AND PAKISTAN UNDER THE NUCLEAR SHADOW

Lt Gen PRAKASH MENON

wisdom tree

First published 2018

ISBN 978-81-8328-524-7

Published by
Wisdom Tree
4779/23, Ansari Road
Darya Ganj, New Delhi-110 002
Ph.: 011-23247966/67/68
wisdomtreebooks@gmail.com

Printed in India

CONTENTS

PREFACE

In September 1998, four months after both India and Pakistan displayed their nuclear prowess, I was in attendance with a well-known Indian Army general at a round-table discussion which was a standard practice that followed lectures of visiting dignitaries at the Defence Services Staff College (DSSC), one of the Indian armed forces' premier training institutions that moved to Wellington from Quetta after Partition. My presence at the round-table was in the capacity of an instructor at the level of a colonel, formally known as the Directing Staff. During the course of the discussion, I discovered that my views on the impact of nuclear weapons on military operations and war as political activity were not only in a minority but it also seemed to be in a minority of one! I felt that though the majority accepted the change, they had not quite appreciated its profundity in its implication and application. This book is a culmination of a long journey that commenced at the conclusion of the round-table, meandered through a PhD thesis and insights gained during my career which culminated as the Military Advisor (2011-14) and Officer on Special Duty (OSD) (2015-17) in the National Security Council Secretariat, New Delhi.

India, after Kargil in 1999, espoused its faith in limited war, a precept not to be confused with limited nuclear war. Rather, it means limited war, which while confined to the conventional realm is restricted in space, time and the quantum of force applied. India's declaration of faith in limited war gave birth to several issues. Is it politically prudent to prosecute a conventional war under the shadow of nuclear weapons? Can such a war achieve substantial political aims? What lessons have been learnt from Kargil, the India–Pakistan crisis of 2002 and several other crises that have ensued? What belief systems should guide the military and political leadership of both countries to minimise the possibility of nuclear war? This book is an attempt to seek answers to the specific questions raised.

The objectives of the book are five-fold. They are:

- To carry out an analysis of the theories of deterrence
- To examine deterrence perspectives in the Kargil Conflict and the India–Pakistan crisis of 2002 and other sporadic crises thereafter
- To examine the theory of limited war and its applicability in the India–Pakistan context
- To study the impact of the conflation of nuclear deterrence and technology-assisted lethality empowerment of state and non-state elements
- To crystallise belief systems that could inform the military and political leaderships

The rationale of the book was derived from the credence of the strategic necessity for investigation and conceptualisation of several issues that arose with India emerging as a nuclear weapons state. It focuses on the potential and limitations of nuclear weapons as it pertains to wars in the India–Pakistan context. My research indicates that in the India–Pakistan context, conventional space below the nuclear threshold is available to prosecute a limited war. But the utility of conventional space and the scope of political objectives achievable through limited war are substantially restricted. Consequently, brinkmanship, punitive strikes, geographically confined conflicts and localised skirmishes are likely to characterise the forms of future India–Pakistan war.

Limited war is possible, and military leaders have no option but to plan for limited war even while being fully aware that escalation cannot always be controlled by operational virtuosity. It is essential that political leaders need to be aware that once you mount the beast you may not know when or where to get off. Worse, one could be reluctantly dragged into the situation despite disinclination. It is an unenviable situation that decision makers on both sides will have to live with as long as both countries possess nuclear weapons.

The research methodology was broadly exploratory and analytical for which an extensive survey of literature was conducted. Reliance was placed on both primary and secondary sources. The primary source of data is through government publications and interview methods. Libraries and internet constituted the main reservoirs of secondary data. In addition, was the availability of newspaper articles and commentaries. I was privileged to enjoy access to the best libraries in the country at the DSSC, United Service Institute (USI), National Defence College (NDC), Institute for Defence Studies and Analyses (IDSA) in Delhi, National Institute of Advanced Studies (NIAS), Bangalore, and University of Madras. Internet also provided unlimited access to a plethora of views on the issues under investigation. I have attempted to acknowledge the sources utilised but in case there are omissions, the fault is entirely mine.

It was my privilege to have had the opportunity to meet and discuss with well-known scholars specialising in the subject which included K Subrahmanyam, Jasjit Singh, VR Raghavan, Satish Nambiar, PR Chari, Stephen Cohen, Scott Sagan, Sumit Ganguly, Ashley Tellis and Bruce Blair, amongst others. Though the views were indeed varied, the exposure was enlightening and they were highly influential in shaping my views.

A special thanks to my PhD Guide and friend, Professor Gopalji Malviya. I must also acknowledge the assistance and advice I received from my colleagues and friends, RK Malhotra, DPK Pillai, Rajesh Barur, Srinath Raghavan, Gurcharan Das, Mekhala and Anjana. I specially thank Joyce and Kannan for their meticulous editing and useful suggestions.

It was during the course of my professional duties as a Brigade Commander that I was involved on the front line in Jammu and Kashmir in the India–Pakistan crisis of 2002 and was also the Major General General Staff (MGGS) at Northern Command during the Mumbai attacks in 2008. Those experiences and several other operational command and staff appointments have certainly influenced my findings on conventional war under a nuclear shadow. My academic pursuits aside, the views expressed in this book have thus been shaped to a significant extent by my professional experiences. Shiela's love, patience, forbearance and quiet encouragement kept the book alive during the course of my long professional journey. To her, my children Rajeshwari, Krishna, Arjun and our grandchild Ayana, I dedicate this book.

INTRODUCTION

> The only salvation is for both countries to follow policies of cooperation and not confrontation.
>
> —General K Sundarji

The advent of nuclear weapons with their awe-inspiring power of destruction has posed several challenges which emanated not only from their destructive power but also from the inability of states to defend against missiles that transported them to targets. Previously, in the conventional era, the failure of deterrence meant an armed clash on the battlefield, resulting in victory, defeat or stalemate with resultant losses and gains serving as a determinant of the outcome. Defence and offensive capabilities served to limit losses and inflict damage upon the opponent.

Nuclear weapons supposedly changed this conventional paradigm by making victory impractical in the context of nuclear-armed adversaries. According to the military strategist Bernard Brodie, 'Thus far, while the chief purpose of the military establishment had been to win wars, after the advent of nuclear weapons, its chief purpose was to avert them'.

Nuclear weapons revolutionised both military strategy and the relationship between force and foreign policy. The targets of nuclear bombs would depend on what each side valued most. States could be dissuaded from going to war, not by the belief that they could not defeat the opponent's military forces, but by the knowledge that they could also suffer unacceptable losses if both have second-strike capability—a capability premised on the notion that opponents could strike back even if they were hit first. It follows, therefore, that unlike the situation in conventional conflicts, absolute power is more important than relative power. In the conventional battlefield, when armies fight armies, knowledge of the absolute size of the forces on either side is somewhat irrelevant because the relative capability of the two forces that are to clash is the key. But with nuclear weapons, the crucial factor is each side's ability to annihilate the other or inflict intolerable damage. For the past three decades, the creation of a missile shield has been suggested as an alternative to being left at the mercy of the adversary.[1] In all likelihood, even if a missile shield becomes practical, countermeasures are likely to render it vulnerable and undependable.

Ever since the advent of nuclear weapons, military planners, strategists and political decision makers have grappled with the possibility that an armed conflict between two nuclear-armed adversaries would degenerate into mutual suicide. This prospect, some believe, kept the peace during the Cold War, and therefore, vindicated the twentieth century American military strategist Bernard Brodie's postulation. The greatest risk posed by war in

[1] Despite enormous amount of time and resources being invested in developing a missile defence shield, it has not been possible for the US, Russia, China, Israel, the United Kingdom and France to develop effective systems that could provide reasonable assurance of protection against their potential nuclear-armed adversaries. India is also in the process of developing a ballistic missile defence capability and has carried out several successful tests so far and claimed that it can be deployed to defend at least two main cities. The main challenge in developing an effective missile defence is the easier ability of the attacker to develop decoys, MIRVs and several other measures that would overwhelm the capability of the defence shield. It could, however, be feasible to develop a system against rogue elements that pose a limited threat.

the nuclear era was the possibility of military application of force becoming detached from achievement of political objectives. Unless the achievement of political objectives remains tethered to rationality and is rooted in costs that one is willing to pay, war can descend into senseless violence. The issue is further compounded by the fact that political objectives, especially in conflict situations, are a product of choices made by policymakers based on internalised belief systems that are often founded on flawed information about the opponent's capabilities, preferences and intentions.

It has been argued that the intensity of political estrangement between the United States (US) and the Soviet Union was insufficient to warrant an armed conflict, and therefore, the belief that nuclear weapons kept up peace may rest on erroneous assumptions. But the lessons of the Berlin Crisis of 1961, Cuban Crisis of 1962 and the US–Soviet nuclear alert during the Arab–Israeli 1973 War and several other near-mishaps and incidents[2] provide a fairly clear picture of the impact of mutual nuclear vulnerability on political decision makers. Precautionary defensive measures, coupled with restraint and flexibility and an overall tendency for caution, had influenced the behaviour of both opponents. In these crises, the consequences of nuclear war weighed heavily on the decisions that were finally arrived at.

The British–American strategic thinker, Colin Gray, said that the inability to actually use nuclear weapons as part of combat does not mean that it cannot be used politically. Instead, it was contended that due to the colossal destructive power of nuclear weapons and their potential to threaten not just the existence of entire societies but even life on earth itself, they could work directly on the source of the real action in world politics—in the minds of policymakers—and therefore, supposedly make greater impact on the realm of

---

[2] Lewis, Patricia; Williams, Heather; Pelopidas, Benoit; and Aghlani, Sasan. *Too Close for Comfort*, Chatham House, 2014. www.chathamhouse.org

grand strategy. Nuclear threats, even those that were conveyed either directly or indirectly, could be straightaway embedded in the psyche of the adversary's political leadership. Fear of suffering damage, in contrast to the ability to inflict it upon the other, could be quickly generated. Therefore, this was the strategic effect that mattered and the rationale for the political utility of nuclear weapons. Ironically, uses of nuclear weapons are also to be found in non-use.

The conundrum created by nuclear weapons was supposedly resolved, albeit questionably, by theories of nuclear deterrence. These theories relied on second-strike capability that assured mutual destruction in case of an attack by one nuclear power on the other. There is no one theory, doctrine or policy of nuclear deterrence, but a whole series of them that differ between countries and between entities within countries.

Nuclear deterrence policies that support only the core function of deterring nuclear threats have much greater credibility than those that seek to use nuclear weapons against conventional and sub-conventional threats. In the dynamics of the India–Pakistan nuclear dyad, the source of instability in the nuclear relationship can be traced to the juxtaposition of Pakistan threatening to use nuclear weapons to deter India's conventional reaction against the use of terrorism as a foreign policy tool.

Consequent to the nuclear tests of May 1998, both India and Pakistan announced their status as nuclear powers. In August 1999, a draft nuclear doctrine was released by India for public debate. The draft embraced the policies of minimum deterrence, no first use (NFU) and civilian control. According to India's nuclear doctrine, the essence of sustaining nuclear deterrence in the Indian context devolves on the achievement and projection of the capability to absorb the first strike from a potential adversary and assure retaliation by residual nuclear weapons that will result in unacceptable damage. In the Indian context, China and Pakistan are considered as the apparent potential adversaries. Geographical proximity between the heartlands of India and Pakistan complicates the deterrent equation. Unlike the bipolar Cold War nuclear situation, India is faced with a tripolar one. These complexities could be compounded if terrorists and non-state actors get access to nuclear weapons, a possibility that is likely in the case of Pakistan. This presents a situation that is unique and unprecedented for India as a nuclear power.

It is debatable whether the India–Pakistan dyad has belied the belief that mutual possession of nuclear weapons deters aggression by conventional or nuclear means with tolerable stability. What has been experienced is that Pakistan has been acting behind the veil of nuclear weapons to intensify its orchestration and support for a proxy war against India through terrorism in Kashmir and the Indian hinterland. This paved the way and resulted in the Kargil Conflict of 1999 and the India–Pakistan crisis of 2002 and several sporadic crises thereafter.

The India–Pakistan situation is also characterised by deep and, thus far, intractable political differences that are further exacerbated by a complex and volatile geopolitical situation involving US, China, Russia, Afghanistan, Iran and Saudi Arabia. Ever since 1998, Pakistan has been emboldened in its prosecution of terrorism and the proxy war unleashed as a significant part of its strategy of 'bleeding India with a thousand cuts'. Pakistan's possession of nuclear capability gave birth to a notion in Pakistan that India would be deterred from retaliating conventionally. Pakistan's faith in nuclear deterrence strengthened over time when they perceived India's reaction to terrorist attacks as hesitant to cross the conventional military threshold.

Pakistan has stated that it would use nuclear weapons in extreme situations, but deliberately prevaricates on the circumstances of use, to both cloak it in ambiguity and thus ostensibly strengthen deterrence. It has claimed that its nuclear weapons neutralise India's conventional superiority and has refused to adopt an 'NFU policy'.

India, on the other hand, has adopted a nuclear doctrine that unequivocally declares that its nuclear weapons are meant to deter only nuclear weapons and other weapons of mass destruction and that it will never use them first. From all indications, both countries have not deployed mated nuclear weapons thus far and may require some preparatory period before their use. However, once India's SSBNs (submersible ship ballistic nuclear) are operationalised, an era of mated systems for the sea-leg of the triad could unfold.

Both countries, however, have acknowledged in the Lahore Declaration of 1999 that 'the nuclear dimension of the security of the two countries adds to their responsibility for avoidance of conflict between the two countries'. Yet, a couple of months later, the Kargil Conflict erupted. A theory of limited war in the India–Pakistan context was thus born and given a well-publicised birth by the Indian defence minister at a seminar in New Delhi in January 2000.

After the Kargil Conflict, three schools of thought emerged in India. Some believed that a conventional war with Pakistan is no longer practical because of the possibility of the war escalating beyond the nuclear threshold, and matches Pakistan's official view. Others are of the view that a war limited in space, time and objectives is possible and Kargil was cited in support of the argument which has also been India's official view. A third group believes that a full-fledged conventional war is possible, because Pakistan's nuclear threats are bluffs that must be called, since Pakistan could well be decimated, even though we could also suffer severe damage.

The idea of limited war is pivoted on the notion that there exists conventional space below the nuclear threshold, which can be utilised to apply military force in pursuance of political objectives. Inherent in the notion was the caveat that only those political objectives, which were perceived as attainable within the ambit of limitations, could be pursued. But since the perception of nuclear threshold was inevitably ambiguous, how much force can be applied, where and for how long, against an adversary who was free to react, was ultimately determined by the ability to take risks, the stakes involved and possibly belong to the realm of the unknown and the inexplicable.

The concept of limited war has also been framed within the realisation that in the nuclear era both adversaries could be annihilated in a nuclear conflict. Hence, any application of military force between two nuclear-armed adversaries carried the potential of escalation into a nuclear conflict. However, it was theorised that it was feasible to control escalation and thus confine the conflict within the conventional realm. The conflict, it was postulated, would be waged in the conventional space below the nuclear threshold. In a fundamental sense, the theory of limited war as hypothesised by India pivoted on the control of escalation based on the belief that nuclear deterrence would endure, as any use of nuclear weapons could only be a decision seeking self-destruction.

Escalation control is likely to remain a contested issue as its reality is concealed in operational details of both militaries that are mostly outside the public domain. It is indeed impossible to predict with any degree of certainty the result of dynamic interaction between opposing military establishments. But more importantly, escalation control has been criticised because it requires an amenable, reasonable, predictable and cooperative adversary; it is one that rarely exists in the practical world and certainly not in the India–Pakistan context.

Escalation of India–Pakistan conflicts into the nuclear realm can never be completely ruled out. Though avoidance of conflicts between nuclear powers is the most ideal method, historical animosities and intensities of political differences between India and Pakistan carry within them the seeds of future conflicts. Pakistan's use of terrorism as a foreign policy tool coupled with its penchant to seek security against retaliation through nuclear weapons is a substantial cause for instability in the India–Pakistan relationship. Future conflicts could, thus, take various forms, under an omnipresent shadow of nuclear war.

A deterrence theory is predicated on the assumption of an adversary rationally weighing the potential costs and benefits of behaviour. Yet, sensible behaviour frequently is in the eye of the beholder and in fact is dependent on the character of the adversary and the context of the engagement. It is difficult to step outside one's own frame of reference to understand that the other side may have very different interpretations of what is considered as sensible and reasonable and consequently, may calculate costs and gains differently. The assumption of a reasonable adversary can lead to mistaken expectations of the adversary's behaviour. Moreover, the feasibility of translating first and second-strike theories into operational practicality has continued to challenge the military leadership. The end result, by and large, has been in the proclivity to over-compensate through larger and larger nuclear arsenals.

An intrusive understanding of deterrence should not be divorced from its context and the opponent. What deters the Chinese may not deter the Pakistanis. Also, what deters Pakistan today may not necessarily deter them tomorrow, especially if there has been a change in the situational context. The differentiated Indian context must not, therefore, be subjected to being viewed through the prism of the Cold War even while nuclear learning is feasible from historical experiences. Application of lessons should privilege the centrality of context.

No nuclear weapons have been employed ever since the use of atomic bombs by the US on Japan in 1945. The context of the use by the US was one of monopolistic possession. However, during the Cold War, the US and Russia accumulated tens of thousands of nuclear weapons and their destructiveness and sophistication grew enormously. Nuclear deterrence in the context of mutual possession, should therefore, be studied from the point of view of the non-use of nuclear weapons.

The lack of an actual campaign involving nuclear weapons and the problems inherent in any attempt to make sense of how such a campaign might develop in the future has not inhibited the development of nuclear strategy. Indeed, the quest for a nuclear strategy that can serve definite political objectives without triggering a holocaust has occupied some of the best minds of our time. By and large, the leaders in this field have been civilians rather than members of the military. This is so because the issues involved relate more to the character of international politics and the nature of higher decision-making in times of extreme crisis in contrast to the employment of force along traditional lines for traditional purposes.

The opportunity for field-testing India's theory of limited war did not take long in coming. The 13 December 2001 attack on the Indian Parliament saw the launch of 'Operation Parakram' by India, who in turn threatened war unless Pakistan withdrew support to cross-border terrorism. The stand-off continued until October 2002 and petered out when India decided to demobilise and resolve issues through the diplomatic route. Limited war, in the India–Pakistan context, remains untested.

The Mumbai terrorist strikes in 2008 did not provoke India into testing its theory of limited war and the military option though examined did not find favour with the political leadership.

Limited war remained a non-event in 2002, 2008 and subsequent crises. The book attempts an examination to understand its potential and limitations through the lens of conventional and nuclear deterrence, while remaining rooted in the dynamics of the India–Pakistan context.

The central argument of the book is that despite their position under the nuclear shadow, India and Pakistan have constructed divergent belief systems regarding the 'modus operandi' of the utility of force to resolve political issues. The strategic approach on both sides is on a perennial collision course, which, while it has no potential to resolve political issues for which force is being applied, serves, instead, only to exacerbate the prevailing difficulties. The political and military leadership needs to adopt an altered mindset that recognises a substantial curtailment of the role of force resulting in it being viewed as more of an instrument of retaliation that can, at best, albeit arguably, assuage heightened domestic emotions. It is now the 'role of force' to make the use of 'force' unattractive to the political and military leadership on both sides.

In an attempt to substantiate the central argument, this study traverses through chapters which include, deterrence policies;[3] the scrutiny of the constellation of forces that pervade the Indian context; the internalised lessons of the 2001-02 Kargil crisis; the Mumbai attack in 2008 and other sporadic crises till early 2017 and their relevance to limited war theories. Such a journey offers abundant evidence to indicate that the ideological pathways espoused by both sides on the utility of force, merely serves to perpetuate their differences. Jettisoning shibboleths and adopting a different strategic approach is an imperative for both countries primarily because the current paths could prove perilous, not only for the subcontinent, but for humanity itself.

[3]The UK deterrent has been considered as mostly an adjunct to the US and hence only the French, China and Pakistan deterrents have been included.

CHAPTER 1

DETERRENCE: AN EXAMINATION

Deterrence is above all a psychological problem. The assessment of risks on which it depends becomes less and less precise in the face of unprecedented novelty and destructiveness. A bluff taken seriously is more useful than a serious threat interpreted as a bluff.

—Henry Kissinger

In common parlance, deterrence suggests either a policy or a situation. A policy of deterrence is usually a calculated attempt to induce an adversary to either do something, or refrain from doing something, by issuing threats primarily in the politico-military and economic activities of world politics. A deterrence situation or system, however, is one where conflict is contained within 'a boundary of threats that are neither executed nor tested; if a threat is executed it is no longer a threat, if it is tested but not executed it is no longer a deterrent'.[1]

Social scientists call deterrence a relational variable. Since deterrence is the product of a relationship, it is not something one

[1] Jones, Roy E. *Nuclear Deterrence: A Short Political Analysis*, Routledge & Kegan Paul, New York, 1968, p l.

can generate unilaterally. Also, it is a shifting value and hence, a variable. Deterrence is a condition wherein the object of deterrent intimidation is induced through threats—not to behave in ways in which one would otherwise have preferred, because of the belief that the consequences must be avoided.

The Phenomenon of Deterrence

At a general level, deterrence is one of the most common modes of behaviour between individuals as well as states. It is neither new to the nuclear age nor pertinent only to military considerations. Indeed, it is a common feature of human relations in society. Deterrence, in fact, is practised widely among family members. Parents are ardent practitioners of deterrence when they try to discourage their children from doing some mischief by warning them of punishment. For parents, deterrence offers real advantages. They are able to shape the child's behaviour as desired through minimum exertion and without open conflict with the child. Moreover, instead of reacting after trouble has occurred, they would have prevented it from occurring in the first place. It is another matter that experience tells us that parental deterrence policies frequently fail! Just as the idea of deterrence is shaped towards maintaining order, it follows a similar pattern in interstate relationships too.

Deterrence is the action of discouraging an action or event through instilling doubt or fear of consequences. It is considered a preferred option rather than dealing with the fallout of the act. Warning signs like 'Beware of dogs', 'Trespassers will be prosecuted' are further examples of the common practice of deterrence. In all the cases, the attempt is to communicate the deterrent message so as to prevent the act from being committed.

The word 'deter' owes its origin to the Latin word *deterrere*, which means to frighten away. The threat of war, open or implied, has always been an instrument of diplomacy by which one state deterred another from doing something of a military or political nature, which the former deemed undesirable. Frequently, the threat was largely latent, the position of the monitoring state being so obvious and strong that no one thought of challenging it. Governments, like individuals, were usually aware of the hazards in provoking powerful neighbours and governed themselves accordingly. Nevertheless, the large number of wars that have occurred in modern times prove that to use force, even what sometimes looked like superior force, has failed to deter.[2] A case in point is the Gulf War, where the US had warned Saddam Hussein that if he used weapons of mass destruction (WMD) or burned Kuwait's oilfields, the US would not hesitate to use nuclear weapons. Saddam did not use WMD but he burnt the oilfields and the US did not respond with the stated threat.

The aim of deterrence is to prevent an enemy from deciding to use armed force; generally, this means compelling it, when faced with a given situation, to act or react to an effective threat. The effect, which it desires to achieve, is, therefore, a psychological one and it is pursued by means of a threat. Inducing fear, and therefore, inaction is the objective. The way in which deterrence acts is basically the exact opposite of war, in that the object of deterrence is to prevent the enemy from taking a decision to act, whereas the object of war is rather to force him to take a decision to accept the conditions imposed on him.[3]

The advent of nuclear weapons, with their awesome power of destruction, did not imbue deterrence with a newer meaning. Instead, deterrence was challenged in its accomplishment if both or all adversaries possessed nuclear weapons and were unable to defend themselves against missiles carrying nuclear weapons. While on the one hand significant destruction could be inflicted on the opponent, on the other, one may also suffer large-scale destruction. Even one

---

[2] Brodie, Bernard. *Strategy in the Missile Age*, Princeton University Press, 1959, p 272.

[3] Beaufre, Andre. *Deterrence and Strategy*, Praeger Publications, New York, 1966, p 24.

Hiroshima-type bomb on a major city would entail destruction that would not be politically acceptable. In the pre-nuclear era, the failure of deterrence meant an armed clash in the battlefield resulting in victory, defeat or stalemate. Losses and gains suffered served as a determinant of the outcome. Defence and offensive capabilities served both to limit losses to oneself and inflict it upon the opponent. Nuclear weapons changed this conventional paradigm by making victory through force application politically unviable in the context of nuclear-armed adversaries.

Contestability

The significance of nuclear weapons is found not only in their high levels of potential destruction or in the speed at which such levels can be achieved, but in a limited range of interactions produced by being capable of achieving high levels of destruction quickly. The combination of rapidity and the scope of their damage means that the costs inflicted on an opponent by nuclear use cannot be effectively contested. This is the fundamental difference between conventional and nuclear warfare. The destruction associated with conventional weapons requires a degree of time and effort that ultimately allows the possibility of adaptation on the part of an opponent. The belief is that one thwart damage that is threatened by the use of conventional forces, as also, the given opportunity to defend is tied directly to the nature of the weapons themselves. Militaries and societies can adapt to the disruption and destruction possible by conventional weapons even at severe levels. By contrast, nuclear weapons are capable of inflicting costs rapidly on such a gigantic scale and with such lasting effect, that little adjustment can be made to blunt these costs to a significant degree. While conventional weapons allow for adaptation to costs, nuclear weapons permit little more than the absorption of punishment. Ultimately, the incontestable nature of nuclear weapons constrains the range of interactions between states in a military deterrence environment and therefore alters the number of options available to states contemplating use of force.

States relying on a strategy of conventional deterrence will tend to have an intense focus on relative power because the retaliatory costs that they portend are ultimately contestable. With conventional deterrence, there exists the potential for finding technical, tactical and operational solutions that may reduce or completely circumvent the threatened level of retaliatory costs. This prospect fuels the intensity of relative calculations since both the states considering the probability of war and one or both also trying to deter it, know they must react quickly to each other's moves. While the price of modern conventional warfare may be extremely high, the concept of winning still endures as reflected in ideas like limited war.

Nuclear Weapons and Statecraft

War, according to the Clausewitzian adage, was merely the continuation of politics by other means. Nuclear weapons made war potentially so mutually destructive, that its application for achievement of political objectives could only be carried out at the risk of one's own destruction. Military victory between two nuclear adversaries, thus, became an impracticable proposition entailing risks that no rational political decision maker would, presumably, ever dare to take. Mutual vulnerability made a crucial difference in how war could be viewed. Throughout recorded history, an all-out war has been a useful tool of statecraft; the ability of states to resort to the highest level of violence has been a tool of international politics. Now that military victory is virtually impossible or entails exorbitant costs and the many forms or outlines that rested on the utility of superior force have been altered, even a substantial military success cannot limit the damage that the other power can inflict. As a result, force and the threat of use of force cannot support foreign policy as they did before the advent of nuclear weapons. But this notion was pivoted on the capability that both sides had invulnerable nuclear forces that would survive the first strike, and yet, inflict, what was termed, 'unacceptable damage' in retaliation. This in effect came to be the leitmotif that provided the intellectual muscle in the formulation of the concept of nuclear deterrence.

The Concept of Nuclear Deterrence

Nuclear weapons provided the ability to directly target population centres by circumventing the conventional military forces available for defence. It was no longer necessary to destroy opposing conventional military capability to decimate the opponent's vital centres of power.

The concept of nuclear deterrence, therefore, relied on the ability of the defender to retaliate with nuclear weapons in spite of having been struck first. It is a capability termed as second-strike capability. The concept could obviously work as long as rationality prevailed on both sides. But if deterrence based on retaliation was to fail, it would not always be rational to carry out the stated threat in a situation of mutual assured destruction (MAD).[4] This was possible because the primary aim of the retaliatory threat was to deter. But if, for some reason, deterrence failed, retaliation would serve little rational purpose, as damage would have already occurred and retaliation may only serve to invite further damage. This internal contradiction, accorded the status of a genuine paradox, undermined the theory. But by invoking 'the threat that leaves something to chance',[5] arguing that a potential aggressor could not count on pure reason to dictate its victim's response, the theory disposed of the conundrum. Rational decision makers realise that the laws of reason are unlikely

---

[4] Steinbruner, John. 'Beyond Rational Deterrence: The Struggle for New Conceptions', *World Politics*, Vol-28, 1976, pp 231-32. Quoted in Blair, Bruce G. *Logic of Accidental Nuclear War*, Brookings, Washington, 1993, p 5.

[5] Schelling, Thomas C. *The Strategy of Conflict*, Harvard University Press, 1981, Chapter 3.

to be obeyed strictly in a nuclear war. Even if circumstances logically compelled a rational victim of nuclear aggression to withhold a retaliatory strike, an aggressor would run the risk of inducing spasmodic retribution. Also, the damage inflicted on the victim's command and control system by the aggressor's attack might, in fact, preclude any chance of rational behaviour.[6]

Securing a second-strike capability in the name of strengthening deterrence powered the debate on nuclear strategy during the Cold War. The operational requirements of a second-strike capability were demanding both in terms of resources and evolution of practicable operational concepts. In practice, it sparked off an arms race that drove the size of arsenals to ridiculous levels.[7] Moreover, with the increasing size and sophistication of the arsenals, both the US and the USSR seriously doubted their capability to absorb first strike and retaliate with substantial effect.

The nuclear posture adopted by both sides to minimise this possibility, was termed as 'launch under attack' or 'launch on warning'; a hybrid between first and second strike that entailed retaliating after the opponent had initiated the first strike but before the warheads had exploded over their targets. This was deemed possible because the missile flight time between the US and the USSR was approximately thirty minutes. Satellite and radar warning was relied upon to give early warning of the attack. During the Cold War and after, fake warnings, due to technical glitches and misinterpretation of natural phenomena, caused false alarms, some of which nearly triggered accidental wars. Despite the end of the Cold War, the hair-trigger posture of that time continues even today, albeit to a lesser extent.

---

[6] Blair, Bruce G. *Logic of Accidental Nuclear War*, Brookings, Washington, 1993, p 5.

[7] At peak levels, both the US and Soviet arsenals consisted of 60,000-80,000 warheads. Cited in 'The Future of the US–Soviet Nuclear Relationship', Executive Summary, CISAC Study, p 36.

Nuclear Strategy

This is fundamentally a form of deterrent strategy that seeks to answer the question 'how to deter?'. Initially, when atom bombs first made their dramatic entrance on the international stage, they were discussed and understood in terms derived from the established theories of air power. Eventually, nuclear weapons became more powerful, more numerous and were in possession by more than one state. New concepts and approaches developed in an attempt to come to terms with a situation in which a war, where the most formidable weapons were used, would in all probability be catastrophic for all concerned. The question that it posed was: whether any useful purpose could be served by the employment of devices that invited discussion–using words like 'holocaust', 'doomsday' and 'Armageddon', and whether any employment could be sufficiently deliberate and controlled to ensure that political objectives were met? Strategy was about the use of force and the threat of force for ends of policy, and nuclear weapons were unusable in a situation where both parties possessed it. What has been at issue has been whether 'nuclear strategy' itself is a contradiction in terms?[8]

The question as to what happens if deterrence fails is vital for the intellectual cohesion and credibility of nuclear strategy. It has not yet been possible to design the means to wage nuclear war that seems a tolerable course of action and that has a realistic chance of leading to a satisfactory outcome. It seems unlikely that such an answer will be devised that does not carry an enormous risk of degeneration into a bloody contest for resolution or a furious exchange of devastating and crippling blows against each other.[9] Thus, nuclear threats and nuclear first strikes are fundamentally implausible, if the parties involved are armed with nuclear weapons. However, some

---

[8] Freedman, Lawrence. *The Evolution of Nuclear Strategy*, St. Martin's Press, New York, 1981, p 18.

[9] Ibid, p 395.

have argued that this does not mean that nuclear weapons have no strategic utility.

The contrary view has been that even if nuclear weapons cannot function as military instruments of victory, like air power, that does not mean that these weapons cannot function strategically. According to this view, while nuclear weapons do not appear to score highly at the operational and tactical levels of military endeavour, nevertheless they can score highly in the realm of grand strategy and policy itself. Nuclear weapons, it is contended, can work directly on the source of the real action in world politics, in the minds of policymakers. To win this 'engagement' is to win the war; thus, they feel, nuclear weapons and Clausewitz are strategically reconciled, even when nuclear war plans seem to offer less assurance. They thus contend that critics of nuclear strategy do not assess the theories at the correct 'level of analysis'.[10]

Some Chinese strategists also reject the notion that nuclear weapons have overturned Clausewitz's axiom that warfare is the continuation of politics.[11] They criticise this argument on descriptive, political and military grounds. Descriptively, the causes of war and its consequences, they say, are not directly linked. The causes of nuclear war exist objectively (interstate conflict, the aggressiveness of hegemonies etc.) and are necessarily political, since the decision to use nuclear weapons would have to be motivated by some political choice or goal. The consequences of nuclear war are a result of such variables as technology, command and control, combat skills and geography. Thus, whether a state can objectively achieve its political goals through nuclear war, and whether a state is motivated by political rationale when using nuclear weapons are two different things. The latter possibility still exists and thus, Clausewitz's axiom

---

[10] Gray, Colin S. *Modern Strategy*, Oxford University Press, 1999, p 323.

[11] Johnston, Alastair Iain. 'China's New Old Thinking', Vol-20, No. 3, *International Security*, 1995-96, p 13.

cannot be overturned.[12] The political grounds are based on the argument that an anti-Clausewitzian view of nuclear weapons leads to a blind opposition of nuclear weapons. This ignores the reality that nuclear weapons exist and complete destruction of nuclear weapons is already impossible.[13] Militarily, the argument is that anti-Clausewitzian views exaggerate the uncontrollability of nuclear war and thus, undermine the credibility of deterrence threats. If the state is incapable of using nuclear weapons short of provoking mutual suicide, then deterrence threats are not credible. The implication is then that credible deterrence at least assumes that nuclear war is controllable and thus, logically there are achievable political goals in nuclear war.[14]

The arguments of the Chinese strategists are disingenuous. In the descriptive argument, they ignore the cause-and-effect relationship that guides decisions. By confining the cause of war to an objective fold, they ignore the weight of consequences on the decision for war. So, it is the rationality of the initiator (the first strike) that cannot conform to the Clausewitzian axiom. The moot question always has been whether rational responses can be expected in the catastrophic circumstances that will inevitably accompany a nuclear strike. To expect it to be so can only be part of a calculated risk in which failure entails horrendous consequences. The political arguments identify a reality but do not address the issue relevant to the Clausewitzian axiom. The military argument seems to aim at building credibility by denying the problems connected with controlling nuclear war. In essence, these are typical issues that arise from notions that try to

[12] Ibid, p 13. Quoted from Diansheng, Fian. 'The invention and use of nuclear weapons cannot change the political nature of war', in NDU Curriculum Research Office, ed, *Essay Series on Military Thought*, NDC Press, Beijing, 1988, pp 47-48.

[13] Ibid, p 3. Quoted from Jianzhi, Zhang. 'Preliminary investigation of questions concerning the basic theory of disarmament', in 'China and the International Disarmament Struggle', *Current Affairs Press*, Beijing, 1987, p 52.

[14] Ibid, p 14. Quoted from Wang and Guo. 'He Zhan Zhengbukenenggaibian', p 43.

threaten a first strike as a tool of policy. In a purely defensive mode, the issue is whether Clausewitz's axiom will still apply and how much can one depend on it?

It is often stated that nuclear weapons are political weapons. But it could be argued that it is impossible to predict the effect of nuclear weapons on the minds of policymakers and fear could result in reactions ranging from paralysis, defiance or irrationality. If the effect is unpredictable, then the end result could even be catastrophic. The risks, therefore, are obvious though unquantifiable. What stake then is worth the risk of nuclear holocaust?

Surely, nuclear weapons should not be allowed to dominate strategy during the course of its conduct and it would be foolhardy to expect the adversary to play the game of nuclear politics by the same rules that one has laid down for oneself.

It is, however, reasonable to assume that the only means to deter the use of nuclear weapons is through the threat of like retaliation. This calls for a second-strike capability that remains invulnerable. This, in turn, is influenced by the military's duty to ensure its survivability under all circumstances. The military's task is challenging, if not conflicting, as it will have to devise plans that, in all probability, cannot deliver a politically acceptable victory. The complications of nuclear strategy are magnified when it is sought to extend the core deterrent function to other areas like deterring conventional aggression or for compellence. At its most fundamental level, it would be foolish to depend on a strategy that is by its very nature engulfed in vast uncertainty at the risk of self-destruction. But nuclear weapons exist and the pursuit of an appropriate strategy, even if elusive, must continue to engage the human mind and spirit.

From an Indian perspective, it would be prudent to analyse the evolution of nuclear doctrines against the background of the contexts in which they evolved. Admittedly, the geopolitical context of the Cold War and the technology levels of the West are different from the existing South Asian context. The commonality lies more in the immutable nature of the colossal destructive capability of nuclear weapons and their powers of deterrence. An examination of nuclear doctrines expounded during the Cold War and after, could, however, provide an insight into the nature of deterrence and its nuances. The French, Chinese and Pakistani nuclear policies are the natural choice due to their greater relevance to the Indian context. The United Kingdom, being closely aligned to the US has been excluded. France's similarity lies in its pursuit of independently deterring a larger adversary.

CHAPTER 2

THE NUCLEAR DOCTRINE OF FRANCE

> The purpose of the French nuclear deterrent was to counter state aggression, wherever it came from and whatever form it may take.
>
> —M Nicolas Sarkozy

The evolution of France as a nuclear power illuminates two domains that are germane to the dynamics of the South Asian subcontinent. France had to deter a stronger power whose power it could not match and it attempted to do it independently, which is not unlike the Sino-Indian dynamic. After the Cold War, France has preferred to maintain 'survivability' as the cardinal virtue and based on its geographical size relied purely on SSBNs. It has steadfastly refused to adopt no first use (NFU) and values its status as an independent nuclear power. Considering similarities of the strategic context, the role envisaged for nuclear weapons is perhaps the key to understanding France as a nuclear power and could provide some guidance in the Indian context.

In the era of the Cold War, France secretly[1] decided to acquire nuclear weapons in 1954 and exploded its first device in Algeria in 1960. French strategic culture and history have been cited as the decisive elements in France's decision to go nuclear.[2] Acquisition of nuclear weapons was perceived as essential to maintaining France's then status as a leading European and world power for three centuries. 'A great state, which does not possess nuclear weapons while others have, does not command its own destiny,' said Charles de Gaulle.[3] The security need arose from France's historical experience of having been invaded thrice in less than eighty years including a humiliating defeat in 1940. The nuclear choice was a 'never again' choice to these memories. Nuclear weapons were, therefore, perceived as a unique tool that provided an autonomous defence policy and facilitated war prevention. The strategic and political value of nuclear weapons was sought to be realised through gaining freedom of action and leverage within the North Atlantic Treaty Organization (NATO) vis-à-vis the US and in the Third World.[4] Some analysts have, however, sought to explain the French arsenal in terms of its symbolic significance rather than the existence of a Soviet threat.[5]

Four Generals (Charles Ailleret, Andre Beaufre, Pierre Gallois and Lucien Poirier) and a civilian strategist Raymond Aaron triggered the early French nuclear debate in the late 1950s and 1960s. These strategists sought to evolve the rationale and means to develop a nuclear arsenal and posture that would deter the Soviet Union.

---

[1] Hymans, Jacques EC. *The Psychology of Nuclear Proliferation*, The Struggle Over the Bomb in the French Fourth Republic, Chapter 4, p 93, https://is.muni.cz/el/1423/jaro2016/MVZ235/um/62791483/Hymans.pdf

[2] Grand, Camille. 'French Nuclear Policy at the End of the Cold War', *Strategic Analysis*, Vol-22, No. 4, IDSA, Taylor & Francis, New Delhi, July 1998, p 528.

[3] Freedman, Lawrence. op cit, p 313. Quoted from Kohl, Wilfrid. *French Nuclear Diplomacy*, Princeton University Press, 1971, p 129.

[4] Ibid, p 528.

[5] Sagan, Scott D. 'Why Do States build Nuclear Weapons', *International Security*, Vol-21, No. 3, Winter 1996-97, p 78.

The French did not publicly acknowledge the Soviet Union as the principal enemy till 1983.[6] What emerged in terms of a nuclear arsenal, the *force de frappe* (strike force) and posture was described as 'deterrence'. The idea was that though France did not have the means to build a huge nuclear arsenal like the US, it would still be able to deter the Soviet Union attacking its 'vital interests' by threatening to inflict an unacceptable level of damage on the Soviet Union, a risk they contended would deter aggression. The interpretation of what constituted 'unacceptable damage' to the Soviet leadership was a matter of political judgment that found reflection in the Presidential choices and targeting guidance.[7]

Gallois was a strong advocate of the idea of 'deterrence by the weak of the strong' and explained the rationale and its requirements. The distinction that he drew between the super power strategies, which, in his opinion, were beyond the range of medium powers as these were based primarily on coercion, and preferred 'dissuasion' as it could be achieved by limited resources and means. It even allows medium powers to deter superpowers from using coercion on them.[8] Since nuclear weapons can inflict immense destruction, a handful of these weapons suffice for a credible deterrent against any threat as long as three conditions are met. The arsenal has to reach a certain level of credibility (ie, a second-strike capability), the vital interests of the 'weak' have to be at stake and the political decision maker should be ready to use the weapons. The second-strike capability was viewed in terms of its proportion to the stake that was sought to be defended.[9] Some have described this as 'proportional deterrence',

---

[6] Ibid, p 9. See also, note 5.

[7] Tertrais, Bruno. *The French Nuclear Deterrent After the Cold War*, RAND, Santa Monica, 1998, p 9.

[8] Grand, Camille. op cit, Note 7, p 543.

[9] Gallois, Gen Pierre. stated, 'The thermonuclear force can be proportional to the value of the stake it was defending', quoted in Kohl, Wilfrid. *French Nuclear Diplomacy*, Princeton University Press, 1971, p 27.

though it has never been part of the official language of the French policy.[10] In doctrinal terms, this meant that the French need not respond to the criteria of MAD: It may be enough to tear off a limb of a potential aggressor, rather than having to threaten his very existence.[11] A principle of reasonable sufficiency was embraced purely on a counter-city strategy. The ability to destroy the adversary's capital was considered a key factor for the French arsenal.[12] In its origin, the strategy was described unashamedly as a 'strategy of the means' (ie, a procurement- and budget-driven strategy). France developed the means it could afford, and then found an intellectual rationale for it. Some logical credibility was sought to be preserved.[13]

The definition of vital interests was a matter of political judgment and has never been clearly defined. The ambiguity regarding vital interests was deliberate. It was felt that nuclear deterrence could cover more ground if the adversary did not know exactly where vital interests began thus rendering the calculation of risk by the adversary difficult. In addition, vital interests, were, by nature, dynamic, in that they evolved over time and were influenced by political, economic, military or cultural parameters.

Another source of deliberate uncertainty was the concept of 'final warning'. This implied that a final warning would be given to any adversary on its way to cross the threshold between non-vital and vital interests. This was sought to be carried out by what was described in 1985 as pre-strategic nuclear weapons. These forces were to deliver a single, final warning blow, leading either to a reestablishment of deterrence or directly to the ultimate strategic

[10] Grand, Camille. op cit, p 529.

[11] Yost, David S. *France's Deterrent Posture and Security in Europe*, Adelphi Papers, No. 194-195, International Institute for Strategic Studies, London, 1985, p 17.

[12] Heisbourg, Francois. op cit, p 303.

[13] Tertrais, Bruno. op cit, p 3.

nuclear exchange.[14] There would be only one massive blow if French vital interests were attacked and the final warning had failed. The French deliberately emphasised the differences between this doctrine and NATO's 'flexible response' to which they never agreed. To the French, the most important threshold was the decision by the adversary to initiate a war and not movement from the conventional to the nuclear realm.

The concept of nuclear battle or war-fighting was rejected due to political and practical reasons. On the political side, it was always judged that signalling a readiness to engage in a nuclear exchange reduced the intrinsic potency of nuclear threats and thus the credibility of deterrence. In reality, France was not considered to have sufficient means for a nuclear battle and, moreover, it lacked the necessary surface area to absorb nuclear strikes and recover from them, unlike the superpowers, which have vast territories.[15] Similar, in fact, is the case with Pakistan.

From the 1980s, the French have emphasised that nuclear weapons are strictly for political purposes only and have referred to nuclear weapons as weapons of non-use that should never be considered for war-fighting. This differentiates them from NATO policies, which they describe as a pure war-fighting doctrine. Yet, they acknowledge that nuclear deterrence without operational capability is not credible. 'Virtual employment' was the term coined by a French strategist to describe the reluctance to envision a failure of deterrence.[16] They have never subscribed to the ideas of multiple strikes and nuclear exchanges, if deterrence failed. For them, deterrence by definition could not fail[17] simply because they abhorred the idea of deterrence failing.

[14] Heisbourg, Francois. 'The British and French Nuclear Forces', Strategic Studies, IISS, London, 1990, p 308.

[15] Tertrais, Bruno. op cit, p 13.

[16] Ibid, p 12.

[17] Ibid, p 13.

During Charles de Gaulle's tenure, an explicitly 'all-azimuths' doctrinal concept was briefly floated after he withdrew forces from the NATO's integrated military command. General Ailleret, a protagonist of the concept, explained its strategic rationale by highlighting the inherent uncertainty in the world order. A capacity was, therefore, required to respond to threats anywhere on the globe. This capability could be acquired through a significant number of megaton ballistic missiles of worldwide range.[18] However, the idea was impractical besides the fact that the French never had the means to achieve such a capability. It was primarily a political statement of independence and aimed to bring about a united Europe at the cost of dismantling the NATO alliance. On the whole, the concept was riddled with contradictions.[19]

The issue of credibility that haunted all nuclear doctrines that espoused deterrence based on first strike also afflicted the French. The credibility of the French nuclear deterrent was undermined by the fact that if the nuclear threat was carried out, it would result in its own devastation. But this weakness was downplayed through the argument that the Soviet Union would be then severely weakened against the US and the UK. This perspective was thought to have a more deterrent effect on Moscow than a tit for tat between a medium power and its much larger aggressor.[20] It is apparent that the credibility and self-deterrence problems inherent in any nuclear strike scenario and its pseudo-resolution by the French also reflected the fundamental limitations and dependency of the French idea as an autonomous arsenal. Because they suspected that the US might not put themselves at risk, de Gaulle's initial rationale for an independent nuclear force was based on a trigger strategy to force the

---

[18] Freedman, Lawrence. *The Evolution of Nuclear Strategy*, St. Martin's Press, New York, 1981, p 322.

[19] Ibid, p 323. See also, Tertrais, Bruno. op cit, p 9.

[20] Heisbourg, Francois. op cit, p 302.

US into a nuclear response in case of Soviet aggression in Europe. The trigger concept lasted till the early 1970s. Thereafter, a process of rapprochement with NATO began.[21]

The French have emphatically rejected NFU as a policy. In their view, it would fundamentally undermine deterrence, might encourage a first strike from the adversary, would stop making major war unthinkable and pave the way for a renewed arms race.[22] They also supported the NATO view that inferiority in conventional forces demanded a 'first-use' policy. The same justification informs Pakistan's nuclear policy.

In the final analysis, throughout the Cold War period, France's nuclear doctrine remained essentially a strategy of Minimum Deterrence that provided very limited options if deterrence failed except as an automatic and full-blown response that was designed to be 'unacceptable'. The credibility and self-deterrence problems remained. Uncertainty was the primary tool utilised to alleviate it and is viewed as an essential element of deterrence. Though France sought to maintain the independence of its arsenal, it could never be totally divorced from NATO in the political and military sense. It, however, gave, France a greater freedom of action in the alliance politics of NATO.

The French nuclear effort was extremely costly in financial terms. At its peak in 1967, it represented 1.2 per cent of the French GDP, 26.4 per cent of the defence budget, and 51.4 per cent of the procurement share of this budget. Altogether, it remained above the 30 per cent mark of procurement for thirty years (from 1963 to 1992) and between 0.4 and 1.2 per cent of GDP per year. This allowed France to deploy an arsenal based on a triad, with nuclear bombers (1964), ground-to-ground ballistic intermediate range missiles (1971 to 1996) and submarine-launched ballistic missile

---

[21] Tertrais, Bruno. op cit, p 17.

[22] Interview with French Foreign Ministry Officials, September 2000.

(SLBMs, 1971). France's nuclear force grew from a few dozen in the 1960s and early 1970s to a few hundred in 1980s, and reached its highest point of 540 warheads in 1991.[23]

The French views on deterring an adversary armed with numerically superior nuclear weapons by threatening to tear off a limb could find applicability in the Sino-Indian context. India could adopt the Beijing criteria and until India develops sufficient missile range capability; the targeted limb could be Shanghai.

Post-Cold War

In the early 1990s, France debated the necessity of a shift in its nuclear doctrine due to the collapse of the Soviet Union and its concerns about nuclear proliferation after the Gulf War. The possibility that some Third World dictators would not be receptive to the language of deterrence was utilised to champion the cause of a more flexible deterrent including lower-yield precision nuclear weapons. The idea was to move from a 'weak to the strong' to 'strong to the crazy' posture. David Cost describes the essential difference in views as being between a 'more operational' approach to deterrence and a 'less operational' approach.[24] President Francois Mitterrand, however, laid to rest any potential move away from the status quo.[25]

The 'French White Paper on Defence', published in 1994, contained some major changes. It put more emphasis on conventional forces and clearly pointed to a reduced role for nuclear weapons. It, nevertheless, acknowledged that though nuclear deterrence remained the cornerstone of the policy, major conventional actions could be undertaken without nuclear deterrence coming into play. This did

[23]Grand, Camille. 'French Nuclear Policy After the Cold War: How to Combine Deterrence and Arms Control', *Strategic Analysis*, Vol-22, No. 4, IDSA, New Delhi, July 1998, p 528.

[24]Yost, David S. op cit, pp 113-139.

[25]Speech of President Mitterrand on the Theme of Dissuasion, 5 May 1994. See also, Tertrais, Bruno. op cit, p 25. See also, Grand, Camille. op cit, p 531.

not mean that France supported the concept of 'conventional deterrence'. Instead 'conventional deterrence' was rejected as an alternative to nuclear deterrence.[26] The French have maintained that their nuclear strategy remains exclusively defensive to be used in extreme circumstances of self-defence, and have stated, 'It could never be an instrument of coercion and even less, a battle tool to be used for military gain'.[27]

In the early 1990s, France expanded the role of its nuclear weapons to include the deterrence of all weapons of mass destruction. In 1995, President Jacques Chirac, in a speech to French ambassadors, said: 'Only the nuclear deterrent force guarantees France against the possible use of weapons of mass destruction, of whatever type they may be.'[28] During the NPT Review and Extension Conference in 1995, France clarified its negative and positive security assurances (NSAs/PSAs). It, however, set specific conditions under which they would apply. First, they were reserved only for those Nuclear Non-Proliferation Treaty (NPT) members that were 'respecting' their commitments. Second, in accordance with Article 51 of the United Nations Charter, France retained its right to self-defence; third, the French deterrent would continue to cover 'any threat against our vital interests and this is valid whatever the means and the origins of the threats, including, of course, that of weapons of mass destruction'.[29] France, thus, utilised a legal principle, known as 'belligerent reprisal'. According to this, violation of a legal norm (the NSA) can be justified if the adversary has first violated another legal norm (NPT or the Biological and Chemical Weapons Conventions). In emphasising the principle of self-defence as an inalienable right,

[26] Tertrais, Bruno. op cit, p 26.

[27] Law No. 96-589 of 2 July 1996 related to military programme for the years 1999-2002, Vol-3, p 12.

[28] Quoted in Tertrais, Bruno. *Nuclear Policies in Europe*, Adelphi Papers No. 327, Oxford University Press for the International Institute for Strategic Studies, London, 1999, p 38.

[29] Ibid, p 38.

France reserved the right of first use in circumstances it deemed as self-defence.

On 9 June 2000, France ratified the Rome Statute of the International Criminal Court (ICC). The instrument of ratification included an interpretative clause that sought to interpret Article 8 of the statute, which named the use of certain categories of weapons as expressly prohibited. France interprets the clause as, 'applicable solely to conventional weapons and can neither regulate nor prohibit the possible use of nuclear weapons nor impair the other rules of international law applicable to other weapons necessary to the exercise by France of its inherent right of self-defence'.[30] It was France's insurance against a future French President being hauled to the ICC for use of nuclear weapons against civilians. France has mostly depended on a counter-city targeting doctrine for nuclear deterrence. But since 2001, France has adapted the term 'centres of power' to indicate its targeting preferences to achieve 'absolutely unacceptable damage'.[31]

During the initial post-Cold War period, France reduced its arsenal from a peak of 540 deployable warheads based on a triad to less than 450 warheads based on a dyad comprising aircraft and submarine-launched missiles. By 2000, the share of the defence budget allocated to nuclear expenditure was reduced by 58 per cent since 1990.[32] The post-Cold War French nuclear consensus has been noted as being built on four pillars—keeping a nuclear deterrent, refusing a war-fighting strategy, Europeanising the French deterrence and promoting disarmament.[33] No fundamental changes to nuclear

---

[30] http://lcnp.org/global/french.htm

[31] 'Centres of power' also understood as 'nerve centres' could indicate a shift away from exclusive counter city targeting. http://dx.doi.org/10.1080/13567888.2015.1024071

[32] Arms Control, Disarmament and Non-proliferation: French Policy, France's Ministry of Defence Document, Paris 2000, p 40.

[33] Grand, Camille. op cit, p 536.

doctrine and strategy have emerged in the post-Cold War period; instead France has played a more active part in discussions on the future of nuclear weapons in international security. It joined the NPT as a nuclear power in 1992 and the Comprehensive Nuclear Test Ban Treaty (CTBT) in 1996.

In March 2008, President Nicolas Sarkozy announced that the French arsenal would be reduced to slightly fewer than 300 warheads. The *Bulletin of the Atomic Scientists* in its Nuclear Notebook of 2013 estimated that France had produced more than 1260 warheads since 1964. The 2013 French White Paper on Defence and National Security[34] stated that the French nuclear deterrent will continue to be fully independent, with the relevant capabilities at the disposal of the President. The nuclear forces will have two clearly separate and complementary components, including the support environment enabling their independent and secure operations. France will continue to sustain these capabilities over time. As a consequence:

- The nuclear ballistic submarine fleet consisting of four submersible ship ballistic nuclear (SSBNs) is being equipped from 2010 onwards with the M-51 intercontinental ballistic missile (approximate range 6,000-8,000 km, each carrying 4-6 warheads), deployed on new generation Triomphant class SSBNs.
- The airborne component has been equipped from 2009 onwards with the Air-Sol Moyenne Portee-A (ASMP-A) cruise missile (500 km range), deployed on Mirage 2000 NK3 and Rafale aircraft, stationed in France and carrier-based. The number of nuclear-capable and land-based aircraft will be reduced from sixty to forty.

---

[34]The French White Paper on Defence and National Security 2013. http://www.defense.gouv.fr/english/content/download/206186/2393586/file/White%20paper%20on%20defense%20% 202013.pdf

- The preservation of national missile and submarine competencies, and the improvement on a 2025 horizon, of the range and accuracy of missiles.

France's decision not to retain land-based missiles could be understood from the limited size of its landmass. The Indian landmass, on the other hand, provides sufficient space for dispersed deployment and should be leveraged for the land-based nuclear forces. The survivability of the Indian arsenal is also enhanced by it having three independent legs. As mentioned earlier, India could in its targeting philosophy adopt the criteria of 'tearing off a limb' when planning deterrence against China. France, when faced with a superior Soviet threat, calculated that the Soviet Union has to preserve the major part of its arsenal against the US. Similarly, while China will always have a larger arsenal than India, it will have to preserve a much larger part of the arsenal to deter the West and Russia. India, therefore, needs to factor in this issue while planning the size of its nuclear arsenal.

CHAPTER 3

NUCLEAR CHINA

> What would we do with so many? To have a few is just fine.
>
> —**Mao Zedong**

India's nuclear capability development was originally animated primarily by the Chinese nuclear threat. While Sino-Indian relations reached its nadir in 1962, the adversarial quality of the relationship has endured. Therefore, in order to seek an understanding not only of how India deals with China but also probe the impact of a deepening Sino-Pak nexus, a study of nuclear China is imperative.

When China exploded a 20 kt fission bomb in October 1964, a statement issued by the government justified the action: 'China cannot remain idle in the face of the ever increasing nuclear threats from the US. China is conducting nuclear tests and developing nuclear weapons under compulsion.'[1] It roundly criticised the Partial Test Ban Treaty (1963) signed by the US, Soviet Union and Britain

---

[1] Statement of the Government of the People's Republic of China, 16 October 1964, quoted in Lewis, John Wilson and Xue, Litai. *China Builds The Bomb*, Stanford University Press, 1988, p 241.

as not only a big fraud but also an attempt to consolidate the nuclear monopoly of the three nuclear powers. It stated categorically that it was China's aim to break the monopoly of the nuclear powers and to eliminate nuclear weapons. The NFU policy was also first articulated in the statement: 'The Chinese Government hereby declares that China will never at any time or under any circumstances be the first to use nuclear weapons.'[2]

The US and other Western powers were characterised as imperialists and deemed to be very arrogant since they possessed nuclear weapons and China did not. But when 'those who oppose them also have such weapons, they will not be so haughty, their policy of nuclear blackmail and nuclear threats will not be so effective, and the possibility of complete prohibition and thorough destruction of nuclear weapons will increase'.[3]

China's Pre-nuclear Experience

There is enough historical evidence to suggest that China was subject to nuclear threats from the US prior to 1964 even though the Sino-Soviet alliance of 1950 supposedly provided China with a nuclear umbrella.

The Korean Conflict was the first such instance. After the Chinese intervention in Korea in late November 1950, at a press conference, the US President Harry S Truman, when specifically asked about the use of the atomic bomb, replied: 'There has always been active consideration of its use. I don't want to see it used'.[4] On further questioning, he said the use of the bomb had always been under consideration as it was one of their weapons. He clarified that whether it would be used against military or civilian targets was for

[2]Ibid, p 242.

[3]Ibid, p 243.

[4]Betts, Richard K. *Nuclear Blackmail and Nuclear Balance,* The Brookings Institution, Washington, 1987, p 33.

the military to decide.[5] The threat, veiled though it might have been, did not work and the Chinese did not hesitate in their offensive.

The second nuclear threat was conveyed during the period of the negotiation impasse prior to the armistice ending the Korean War in July 1953. The US President Dwight D Eisenhower, in his memoirs, described the nuclear aspects of the threat as only thinly veiled: 'To let the communist authorities understand that, in the absence of satisfactory progress in negotiations, we intended to move decisively without inhibitions on our use of weapons'.[6] The Chinese, however, perceived the effect differently. According to Jiang, who served with the Chinese delegation at the Panmunjom talks from 1951 to 1953, Beijing had resolved to stand firm in the face of American blackmail.[7] What remains undisputed is that the Chinese were subjected to nuclear threat.

The third experience of nuclear threat was during the Quemoy and Matsu crisis that was initiated by the Chinese in September 1954, when the Chinese artillery launched a bombardment of the Nationalist-held outpost on the Island of Quemoy. Two American soldiers were killed. This was a sequel to several previous incidents that included the hijacking of ships trading between China and Taiwan, patrolling of the US carrier-based aircraft up to the international boundary close to the East China coast, and increased overflights by Nationalist aircraft. In August 1954, the Nationalist Navy reportedly engaged and sank eight Chinese gunboats. The Chinese had accidentally brought down a British airliner, apologised and offered to pay reparations. During the rescue operations following the incident, a US carrier-based aircraft shot down two Chinese patrol fighters and strafed Chinese gunboats.[8]

---

[5] Ibid, p 33.

[6] Newton, Jim. *Eisenhower: The White House Years*, Vol-1, Doubleday 1963, p 181.

[7] Lewis, John Wilson and Xue, Litai. op cit, p 15.

[8] Ibid, p 25.

Two months later, the Tachen Islands further to the north of Matsu and Quemoy were bombed. One hundred Chinese planes raided the Tachens in January 1955 and a week later they seized Ichiang Island just north of the Tachens. In February, the Nationalists were evacuated from the Tachens with American assistance.[9] It was the evacuation of Tachens that brought home to the Americans the realisation that 'massive retaliation' was an impotent strategy.[10] Clearly, the stakes at issue could not find political justification for massive use of nuclear weapons.

The defence of Quemoy was perceived as requiring the use of nuclear weapons. Eisenhower and John Foster Dulles made a number of public statements, leaving little room for doubt that nuclear detonations would be part of the military resistance to a Chinese invasion. In early January 1955, Eisenhower hinted to reporters that escalation would be an inevitable consequence of military engagement.[11] In February 1955, Eisenhower ordered a series of publicly announced tests for tactical nuclear weapons (TNW). In March, Eisenhower commented to a reporter, 'In any combat where TNWs can be used on strictly military targets and for strictly military purposes, I see no reason why they shouldn't be used just as they would use a bullet or anything else'. In his memoirs, he admitted that he expected the statement to have some effect in persuading the Chinese about American determination.[12] The Chinese reacted through a Radio Beijing broadcast that accused Eisenhower of purveying the falsehood that nuclear weapons could be employed without massacring civilians.[13] In late April, the Chinese announced

---

[9] Betts, Richard. op cit, p 54 and Lewis, John Wilson and Xue, Litai, op cit, p 15.

[10] Lewis, John Wilson and Xue, Litai. op cit, p 32.

[11] Eisenhower, Dwight D. 'Public Papers of the President of the United States: Dwight D Eisenhower', US Govt. Print Off., 1955. Quoted in Betts, Richard, op cit, p 59.

[12] Eisenhower, Dwight D. op cit, p 477.

[13] Betts, Richard. op cit, p 59. Quoted from Hoopes, Townsend. *The Devil and John Foster Dulles*, Little, Brown & Co., Boston, 1973, p 279.

their willingness to negotiate. This decision could be attributed speculatively to the American nuclear threat.

The second Quemoy crisis was sparked by China blockading the islands, accompanied by artillery and air strikes in August 1958. In an attempt at extended deterrence, the Soviet Union warned that a threat against the Chinese would constitute a threat against the Soviets. The US publicly stated that it was ready to use nuclear weapons to defend Quemoy. By the end of September, the blockade was broken and the crisis eased. The US, however, transferred to Quemoy eight-inch Howitzers that were nuclear-capable. Though no nuclear ordinance was deployed, the use of nuclear intimidation was obvious. By early September, the Chinese offered to resume negotiations. In 1963, the Chinese in an official statement debunked the notion that Soviet nuclear weapons played any influential role in the 1958 crisis. They postulated that Soviet declarations of support came after the Soviets were convinced that there was no possibility of nuclear war.[14]

How much the Chinese were deterred from pursuing the takeover of the islands cannot be conclusively surmised. But there was certainly no real intention nor capability of the Chinese to attempt an invasion of Taiwan, as it was made out to be, in order to invoke the US involvement based on the US–Taiwan Defence Treaty[15] signed on 5 December 1954. The US committed itself to oppose an armed attack on Taiwan, the Pescadores and 'such other territories as may be determined by mutual agreement'.[16] But it would be reasonable to assume that the futility of pursuing force in a situation of nuclear asymmetry would have been obvious to the Chinese. The lesson China learnt from the Korean and the Taiwanese experience

---

[14] Betts, Richard. op cit, p 75.

[15] Ibid, p 30.

[16] Shuguang, Zhang. Deterrence and Strategic Culture: Chinese-American Confrontations, 1949-58, p 212.

was that as long as nuclear weapons asymmetry existed it would be vulnerable to blackmail. Also, it would be deterred from changing the status quo through the use of force. In a nuclear confrontation, one could not rely on another nation for protection. To them, the notion of extended deterrence was an anathema. These lessons find clear expressions in the Chinese statement after the 1964 blast.

Nuclear Rationale

The Chinese decision to acquire nuclear weapons was supposedly taken in January 1955.[17] It is quite obvious that the 1954 Quemoy-Matsu Crisis and the American nuclear threats were instrumental in that decision. Chairman Mao stated at a Politburo meeting in April 1956, 'In today's world, if you don't want to be bullied by others, we should have atomic bombs by all means.'[18] China's refusal to accept India's nuclear weapons despite unresolved territorial disputes and the volatile political relations with China and Pakistan were, therefore, hypocritical.

Ever since China developed its nuclear arsenal, it has not been subject to nuclear blackmail or threat. How much of this is attributable to the possession of nuclear weapons is impossible to determine. But the absence of threat is an incontrovertible fact. In 1995, a *New York Times* report regarding a statement made by a Chinese general visiting the US was portrayed as an implied nuclear threat. The incident came to be known as the Los Angeles quote. During a discussion with the US Ambassador, Charles W Freeman regarding the American reaction to Chinese military manoeuvres in the Taiwan Straits, the Chinese were justifying their view that there would be no American military reaction. At the end of a rather heated discussion, one of the Chinese generals said: 'And finally you do not have the strategic leverage that you have had in the 1950s when you

---

[17] Ibid, p 11.

[18] Shuguang, Zhang, op cit, p 281.

threatened nuclear strikes on us. You were able to do that because we could not hit back. So, you will not make those threats. In the end, you care more about Los Angeles than you do about Taipei.'[19] This elaboration is convincing in the sense that it was a statement in the context of deterrence and not a threat to bomb Los Angeles. But the statement of the Chinese general highlights the Chinese belief that it is their possession of nuclear weapons that has kept nuclear blackmail and threat at bay and a threat to Los Angeles could be suggestive of the idea of 'tearing off a limb'.

Nuclear Policy Guidance

Till the early 1980s, China had never articulated any nuclear policy except a policy of NFU. For decades, Chinese leaders often stated that China possessed nuclear weapons and their delivery means in order to prevent blackmail and coercion by the other nuclear powers, principally the US and the then Soviet Union. This statement combined with the small and the relatively unsophisticated nature of China's nuclear force structure had led most analysts to conclude that China subscribed to a policy of minimum deterrence which relied on countervalue targeting. As a representative of China's views on this issue, in 1983 Deng Xiaoping stated, 'China only wants to adhere to this principle: We must have what others have, and anyone who wants to destroy us will be subject to retaliation.'[20] But the underlying causes of China's unsophisticated and relatively small arsenal at that time can probably be attributed to its internal political upheavals and lack of economic and technological means.

Some of the major policy declarations of China after the 1980s have been:[21]

---

[19] Proliferation Brief: Vol-2, No. 10, 11 May 1999. https://www.armscontrolwonk.com/archive/200433/gertz-and-xiong-a-love-torn-asunder/

[20] Quoted in Lewis, John Wilson and Litai. China's Strategic Sea Power, Stanford University Press, 1995, p 233.

[21] Ibid.

- China's nuclear weapons are purely for defensive purposes.
- Since 1964, China has adopted a universal NFU pledge. (China applies NFU to Taiwan.)
- China provides non-nuclear weapon states with unconditional negative security assurances.
- China participates in several nuclear weapon-free zone treaties in Latin America, the South Pacific, Southeast Asia and Africa; these commitments prohibit China from deploying, using or threatening to use nuclear weapons in these regions.
- China has pledged not to target its nuclear weapons against the US or Russia.
- China has urged the US and Russia to make deep cuts in their nuclear forces, and advocates the complete elimination of nuclear weapons.
- China opposes the development and deployment of space-based weapons and missile defences.
- China supports an early conclusion of a treaty banning the production of fissile material. (It has, however, made no official declaration regarding stoppage of fissile material production.)
- China has agreed to a moratorium on nuclear testing and has signed but not ratified the CTBT.

The trajectory in the development of China's nuclear weapons programme has been guided by an indigenous thought process that supposedly differs substantially from the Western notions. Notably China's view on nuclear policy has been influenced by its political leaders with the military playing, in the most part, a subordinate role. Nuclear deterrence in their view is to be restricted to the core role of deterring nuclear attacks. They consider that countering coercion rather than nuclear deterrence is the central role. In fact, the Chinese translation of deterrence is '*weishe*' which means coercion. They consider nuclear coercion a much more realistic threat than a nuclear attack.

Based on an understanding that first use of nuclear weapons is not a choice in the real world, they do not subscribe to 'nuclear deterrence based on first use'. While a demonstration of retaliatory nuclear capability is required, it does not require large numbers nor does it have to be permanently on alert. China believes that a small and survivable nuclear force in a de-alerted and de-mated mode is adequate for the purpose of countering coercion.[22] Therefore, no quantitative competition resulting in an arms race is justifiable. China, however, views technical lagging as a major security challenge that could provide the opportunity and invite an attack in a crisis. China has, therefore, reacted to the US development of ballistic missile defence (BMD), surveillance, precision strike and counterforce capabilities like multiple independent reentry vehicles (MIRVs).

Unlike the US and Russia, China has not allowed itself to be persuaded by the threat of a surprise 'bolt from the blue' attack. It has continued to maintain a de-mated posture and its nuclear force structure and deployment plans are crafted to speedily mate and be ready for counter-retaliation when ordered. This has also been India's methodology. If China changes its de-mated posture, it could, in all probability be attributed to a deepened influence of the military viewpoint in its nuclear policy.

In 1996, the American magazine, *Newsweek*, reported that Sha Zukang, China's arms control negotiator, in an interview reportedly stated that China's commitment to a 'no-first-use' of nuclear weapons against any state did not extend to Taiwan. This received wide publicity but the clarification issued subsequently did not. Issued by a Chinese foreign ministry official, it explained that Sha's statement was in the context of NFU policy being inapplicable to Taiwan as a state since it was Chinese territory. He reaffirmed that NFU was applicable to Taiwan.[23]

---

[22] Li Bin, 'China's potential to contribute to multilateral nuclear disarmament', *Arms Control Today*, Vol-41, No. 2 (March 2011), pp 17-21.

[23] Agence France Presse, *Dateline Taipei*, 6 August 1996.

In a news briefing on 2 September 1999, the Ministry of Foreign Affairs Spokesman Sun Yuxi, said China would not use nuclear weapons against Taiwan. 'We will not be the first to use nuclear weapons, we will not use nuclear weapons against non-nuclear weapon countries and we will not use our nuclear weapons against our Taiwan compatriots,' Dow Jones quoted him as saying.[24] Some Indian analysts, probably ignorant of the clarifications, have transposed Sha's statement about Taiwan and NFU to Arunachal Pradesh and expressed reservations about China's pledge.[25] On 24 March 2004, the author posed a similar query to the Chinese Defence Minister General Cao Gang Chung, at an interactive session in Delhi. He categorically denied that the Chinese NFU pledge had any exceptions. In 1994, China had submitted a draft treaty on NFU of nuclear weapons to the Conference on Disarmament at Geneva. In April 2009, Chinese Foreign Minister Yang Jiechi in his address to the conference again called on the other nuclear weapon states to follow China's example and announce a blanket NFU doctrine.

In July 2000, during the Central Military Commission conference on strategic military equipment, President Jiang Zemin outlined the 'five musts' on nuclear weapons:[26]

- China must own strategic nuclear weapons of a definite quality and quantity in order to ensure national security.
- China must guarantee the safety of strategic nuclear bases and prevent against the loss of combat effectiveness from attacks and destruction by hostile countries.
- China must ensure that its strategic nuclear weapons are at a high degree of war preparedness.

---

[24] Gesteland, Lester J. reporting for Chinaonline.com, 2 September 1999.

[25] Singh, Swaran. *China's Changing National Security Doctrine, USI Journal,* Institute for Defence Studies and Analyses, 1999, p 237.

[26] http//:www.cns.miss.edu/research/china/index.htm

- When an aggressor launches a nuclear attack against China, China must be able to launch nuclear counter-attack and nuclear reattack against the aggressor.
- China must pay attention to the global situation of strategic balance and stability and, when there are changes in the situation, adjust its strategic nuclear weapon development strategy in a timely manner.

The five 'musts' reflect the thrust of the Chinese nuclear modernisation programmes. A major change in the strategic situation probably alludes to the US intentions to deploy the National Missile Defence and Theatre Missile Defence systems. There has been speculation among analysts that China is moving from a minimum deterrence doctrine to one of limited deterrence.[27] This speculation has been based on the writings of some Chinese strategists who have found minimum deterrence inadequate to confront a situation that commences with a limited strike. Instead, limited deterrence calls for counterforce and countervalue, tactical, theatrical and strategic nuclear forces to deter the escalation of conventional or nuclear war. They feel that if deterrence fails, this capability should be sufficient to control escalation and to compel the enemy to back down.

The development of a Chinese capability aimed at resolving the familiar problems that haunted the development of the US and the Soviet Union's arsenals will probably set them on the same track. There is obvious merit in the argument that minimum deterrence capabilities are vulnerable to a first strike from an adversary with a much larger arsenal and are incapable of controlling any escalatory competition or achieving intra-war deterrence. But this is primarily a military argument. In seeking technological fixes to what is fundamentally a political issue, China may build arsenals that will still suffer from the fundamental problem of credibility which afflicts

[27]Johnston, Alastair Iain. 'China's New Old Thinking', Vol-20, No. 3, *International Security*, 1995-96, pp 5-42.

any nation declaring a first-strike policy. But as the Chinese are strong proponents of the NFU, so the move to limited deterrence, if at all it takes place, will probably reflect the gains achieved by the military viewpoints in political decision-making.

Nuclear Force Development

China has in recent decades developed shorter-range nuclear forces capable of reaching Japan, Taiwan, India and eastern part of Russia. The DF-21 and DF-21A are the primary regional nuclear missiles with a range of 1750 and 2,150 km and a warhead of 200-300 kt. Western intelligence sources estimated in 2016 that China's inventory of DF-21s also includes a third nuclear version apart from two conventional versions: DF-21C land attack missile and the DF-21D anti-ship missile. The nuclear versions are estimated to have 80 launchers. Some of the DF-21 missiles are deployed in an India-specific configuration at Delingha in Qinghai province, from where the missile which is rail and road mobile could be rapidly moved to forward locations in Tibet via the Qinghai Tibet Railway (QTR). The QTR has several tunnels some of which are believed to have been looped which could facilitate mobile-cum-hardened deployment of intermediate-range ballistic missile (IRBM) assets. The mix of nuclear and conventional DF-21s impacts stability in a crisis situation as it carries the risk for misunderstanding, miscalculation and therefore, escalation.[28]

Missile developments and overall military modernisation have been accompanied by substantial progress in China's space, cyber and electromagnetic domains capabilities. In 2016, China established the Strategic Support Force (SSF) to provide information support. The SSF is expected to consolidate control of People's Liberation Army's (PLA)

---

[28] Kristensen, Hans M and Norris, Robert S. 'Chinese Nuclear Forces 2016', *Bulletin of the Atomic Scientists*, Nuclear Notebook, 72:4, 205-2011, DO I:1080/00963402.2016,1194054

space-based C4ISR (command, control, communications, computers and intelligence, surveillance and reconnaissance) capabilities.[29]

The Chinese have never acknowledged the existence of tactical nuclear weapons (TNWs). In the 1990s, some sources credited the Chinese with possessing about 150 of them.[30] China's 150 TNWs are estimated to include artillery shells, atomic demolition munitions and short-range missiles.[31] Later, these sources admitted that their estimates were speculative. The TNWs were supposedly developed in the 1970s to be used to stop a Soviet armoured blitzkrieg across the northern and western borders.[32] Since the decline of the Soviet threat in 1980s, some Chinese strategists have hinted that nuclear weapons in a theatre of war role will be required to fight limited wars of the future. But without any articulated official policy, the doctrinal prescriptions that underlie Chinese tactical weapons will remain within the realm of conjecture. In 1999, in response to a US congressional report that China had stolen US nuclear and missile secrets, China issued an official rebuttal that stated: 'China had no other choice but to continue to carry out research and development of nuclear weapons technology and improve its nuclear weapons systems, mastering in succession the neutron bomb technology and the nuclear weapon miniaturisation technology…through its own efforts over a reasonable period of time'.[33] The fact that the Chinese

---

[29] Elsa Kania, PLA Strategic Support Force: The "information umbrella" for China's military, *The Diplomat*, 1 April 2017, https://thediplomat.com/2017/04/pla-strategic-support-force-the-information-umbrella-for-chinas-military/

[30] http//www.bullatonsci.org/issues/nukenotes/nukenote.html

[31] Saunders, Philip C and Yuan, Jing-dong. 'China's Strategic Force Modernization', Centre for Non-proliferation Studies. http//:www.cns.edu/research/china/index.htm. See also www.bullatomsci.org/issues/nukenotes/nukenote.html

[32] Johnston, Alastair Iain. op cit, p 27.

[33] Ferguson, Charles D; Madeiros, Evan S and Saunders, Philip C. 'Chinese Tactical Nuclear Weapons', quoted in Millar, Alistair and Alexander, Brian. *Tactical Nuclear Weapons: Emergent Threats in an Evolving Security Environment*, Potomac Books, Virginia, 2003.

possess the capability to develop TNWs is not in doubt, but what is unknown is the extent to which such capability has been utilised to develop and deploy TNWs. Most reports of China possessing these weapons are either unconfirmed Western intelligence reports or guesstimates of think tanks. For China, such uncertainty serves the objective of increasing ambiguity and strengthening deterrence at the sub-strategic level. It is also possible that China has given TNW technology to Pakistan, given existing claims of developing TNWs.

For the past several decades, China has maintained an active nuclear weapon and missile development programme. Estimates of the Chinese nuclear arsenal vary substantially depending on the source. The US Defence Intelligence Agency (DIA) estimated in 2013 that China deployed 300 nuclear warheads for land-based, silo-launched ballistic missiles, submarine-launched missiles, mobile missiles and aircraft-delivered free fall bombs.[34] In 2016, the *Bulletin of the Atomic Scientists*, Nuclear Notebook, stated that China had a total inventory of approximately 260 warheads including those produced for SLBMs or awaiting dismantlement.[35] The Nuclear Notebook's basing of the warheads largely conforms to the DIA estimates. The land-based missiles include several types and its variants: DF-4, DF-5A, DF-5B, DF-21, DF-31 and DF-31A. The DF-5A is silo-based and the rest are mobile. China's land-based Dongfeng (DF) series of missiles range from 1,800 km (DF-21A) to 13,000 km (DF-5A ICBM).

Even though China refrained from deploying multiple warheads for several decades despite having the capability, it appears to have commenced deployment as a reaction to the ballistic missile systems of the US. In 2015, it was estimated that about half the 20 DF-5B had been equipped with three MIRVs each. In addition, the

[34]Strategic Weapons Systems, China, *IHS Jane's Defence Weekly*, 1 July 2013, IHS Global limited.

[35]Kristensen, Hans M and Norris, Robert S, op cit.

DF-41 ICBM had been under development for several decades and was reported to be capable of carrying MIRVs and penetration aids to counter the US ballistic missile system.[36] China is also in an advanced stage of developing hyper glide vehicles (HGV) DF-ZF. By November 2015, it had carried out six tests and claimed to have achieved a speed of Mach 10. The HGV warhead glides in a flat trajectory after reaching an altitude of 100 km.[37] It cannot be intercepted by the existing BMD systems that rely on satellite-based sensors along with ground and sea-based radars that are designed to intercept missiles with a predictable ballistic trajectory. China could utilise its HGV capability for delivery of conventional and nuclear warheads.

China is also in the process of replacing its liquid-fuelled missiles with solid-fuelled ones. One of the major advantages of this move is the increase in survivability and retaliation due to mobility and decreased time taken for preparations for the firing mode. The newer generation DF-31A missile, a solid fuelled, three-stage, road mobile intercontinental ballistic missile (ICBM) with a 11,200-km range and a payload of 200-300 kt, has been introduced in service. It is not clear whether it will replace the older DF-5A as the DF-31A's range and payload are less than the DF-5A's (13,000 km and 4-5 mt). The DF-4 ICBM (5,400 km and 3.3 mt) are being replaced by the DF-21 (2,150 km and 200-300 kt) and DF-31 (7,200 km and 200-300 kt).

The inventory of missiles includes ICBMs, medium-range ballistic missiles (MRBMs), short-range ballistic missiles (SRBMs) and land attack cruise missiles (LACMs). Their basing is a combination of silos, tunnels, rail and road mobility. Some of the older ICBMs like the DF-5As are silo-based. However, the newer missiles like the

[36] Ibid.

[37] Franz-stefangady. China tests new hypersonic weapon, *The Diplomat*, 26 November 2015. http://thediplomat.com/tag/wu-14-hypersonic-glider-vehicle-hgv

DF-21, DF-15, DF-11 and DF-31 are road or rail-mobile. They are likely to be towed to pre-surveyed launch locations and being solid-fuelled can be launched rapidly. A major portion of the ICBMs are based to cover the US, Europe, Russia, India, East and Southeast Asia. The larger portion of the MRBMs and SRBMs are based to cover Taiwan and Japan. The DF-21s, based in Yunnan, Qinghai and Xinjiang, cover India and are the primary regional nuclear missiles that have seen maximum expansion in numbers in the last decade.

Besides upgrading its land-based missiles, China is also developing its submarine-based capability. China's Type 092 SSBN with 12 JL-1 SLBMs (Submarine Launched Ballistic Missiles), (1,000+ km, 200-300 kt) became operational in 2003.[38] Though China developed its second-generation new Jin class SSBNs (type 094) in 2004, it has not been successful in operationalising its JL-2 SLBM which is still undergoing flight-testing. The JL-2 is a modified version of the DF-31 with a range of 7,200 km.

By 2016, China was reported to have four types of 094 SSBNs with one more under construction. Unconfirmed reports suggest that future 094 SSBNs could carry up to sixteen or twenty-four SLBMs. Once operational, the JL-2 will provide a submarine-launched capability, from waters near China to target India, Hawaii and Russia and western parts of the US. There is doubt whether the Jin class has ever conducted deterrent patrols against the US as it would entail sailing to the Pacific Ocean, through choke points would thus be vulnerable to anti-submarine warfare.

Command and Control

China's nuclear forces are under strict political control. All ballistic missile and nuclear submarine missile forces are under the direct command of the Central Military Commission (CMC). The Second

[38] Strategic Weapon Systems, China, *IHS Jane's Defence Weekly*, op cit.

Artillery Corps, designated in 2016 as the PLA Rocket Forces (PLARF) is the operational structure that houses China's nuclear and conventional missile capabilities. It comes under the operational control of the military supreme command or general staff department (GSD).

China is unique amongst nuclear powers that have its conventional and nuclear missiles under the same command structure. There has been no explanation for such an arrangement. The major impact is on crisis stability, as it is not possible to distinguish between nuclear and conventional missiles, except that conventional missiles could be mostly confined to shorter ranges in comparison to nuclear missiles. But if conventional missiles are part of an anti-access and anti-denial strategy pertaining to the Ocean space, one could expect conventional missile with longer ranges. A possible reason for keeping both under a single structural roof could be to project an inflated quantity of nuclear missiles by leveraging the difficulty of distinguishing between the two. Visible, common signature facilities and equipment like launch pads and vehicles could make it easy to conceal the actual numbers of nuclear missiles and also portray greater numbers.

Missile Defences

China has been developing a missile defence for several decades. On 10 January 2010, China carried out its first ground-based mid-course interception test. China became the only country after the US to demonstrate the interception of a ballistic missile with a kinetic kill vehicle. This was followed on 27 January 2013 with a second ground-based, mid-course interception test. It could be assumed that China is moving forward steadily towards an operational BMD capability. Some sources indicate that China could deploy the BMD by 2025.[39]

---

[39] Strategic Weapons Systems, China, *IHS Jane's Defence Weekly*, op cit.

China acquired from Russia the S-300 surface-to-air missiles (SAMs) in the 1990s and it tested the new Russian S-300PMU-2 SAMs in 2006 against its DF-15 and DF-11 SRBMs. The Russian system could not intercept the Chinese SRBMs. By 2009, China was supposed to have developed a 125-km range FD-2000 anti-tactical ballistic missile (ATBM) which had been tested in the anti-missile mode. China is also learnt to be developing 400-km range ATBMs.

It is unlikely that China's BMD will be able to provide any reasonable degree of assurance against the concurrent development, by other powers, of countermeasures to overcome the system.

Force Modernisation Trajectory

The future form and content for the modernisation of the Chinese arsenal will be mainly driven by the state of Sino-American relations, the impact of the US arms supply to Taiwan and Japan, the deployment of the US National Missile Defence, Theatre Missile Defence systems and Prompt Global Strike capabilities. However, unforeseen developments globally like South China Sea disputes, North Korea and President Trump's exertions could also alter the existing trajectories. Its pace is likely to be governed by economic factors and the development of indigenous technology. India must view the Chinese arsenal as a threat which is posed by its mere existence. In the long term, as long as Sino-Indian disputes endure, an unknowable portion of the Chinese arsenal can be directed at India. Such a threat can be countered with a policy of minimum deterrence and second-strike capability to cause unacceptable damage based on the concept of 'tearing off a limb'. The fact that China can never afford to expend all its arsenal only on India alone should restrain the proclivity to base decisions on quantitative comparisons. Instead survivability and retaliation should be the guiding beacon for India. Moreover, China's nuclear modernisation is primarily to increase its second-strike capability against the US. The shadows cast by the

Sino-US competition should not be the main driver in developing India's nuclear capability even though some effects are unavoidable. Development of limited BMD systems, MIRVs and HGVs by India may seem inevitable but should be assessed for its consequences for stability and arms race.

Sino-Pak Nexus

China has for several decades persisted in nuclear and missile assistance to Pakistan in order to keep India tied down to the subcontinent. It is no coincidence that Indian technological breakthroughs in nuclear and missile technology were accompanied by Chinese transfers of similar technology to Pakistan. Some examples are given below:

- 1974 Indian Pokhran nuclear explosion—1976 China-Pakistan nuclear cooperation agreement.
- 1983 IGMDP announced by India—In 1984, China supplied nuclear weapon design to Pakistan.
- 1988 Prithvi test—The same year, Chinese assistance to Pakistan's Hatf missile programme begins.
- 1990 Agni test—Between 1991 and 1994, eighty-four M-11 missiles supplied to Pakistan.
- May 1998 Indian nuclear tests—China provided Pakistan assistance to conduct tests including supply of cables and connectors.

China's nuclear and missile proliferation has been documented in detail and continues apace albeit clandestinely. A September 2016 report entitled 'Alpha Report'[40] illuminates the Sino-Pak nexus. This is despite China's commitment to international proliferation agreements. Notably, the international community has done very little to hold China to account.

---

[40] Alpha Report, Centre for Science and Security Studies, Kings College, London, September 2016. http://projectalpha.eu/wp-content/uploads/sites/21/2016/11/20160929-pakistan-public-version.pdf

SINO-INDIAN NUCLEAR DYNAMICS

China's reaction to development of US technological capabilities and proliferation of nuclear and missile capabilities to Pakistan could find resonance in development of India's capabilities. But essential Chinese[41] and Indian approach to nuclear weapons could be expected to dampen down such pressures.

The Sino-Indian border dispute and regular face-offs are unlikely to provide any role for nuclear weapons unless of course a major war transpires either deliberately or unintentionally. Alerting of nuclear weapons must be treated as the red line that must be avoided.

There is an unmistakable similarity between the Chinese and Indian arguments for developing nuclear weapons. But since India remains outside the NPT, China has so far been disinclined to discuss nuclear issues bilaterally. There is also congruence of views on the role of nuclear weapons, NFU, minimum deterrence and de-mated deployment. These convergences could be diplomatically pursued towards reducing the role of nuclear weapons through a universal NFU treaty and subsequent global disarmament.

The role of nuclear weapons in a future Sino-Indian conflict is likely to be circumscribed by two main issues. Firstly, both nations have a NFU policy that should curtail any proclivity to indulge in coercion through nuclear threats. Secondly, the vital issues at stake will be the main drivers in regard to the scope of the conflict and the determining of the possibility of nuclear weapons coming into play. The scale of conflict in military terms will be a primary determinant which through inadvertence, miscalculation, misjudgement and misperception could dictate the entry of nuclear weapons into the arena. The issue is compounded by the fact that China houses both conventional and nuclear weapons in its PLARF. This, therefore, increases the chances of an Indian misperception and misjudgement

[41] Lewis, Jeffrey. *The Minimum Means of Reprisal*, The MIT Press, 2007.

during actual hostilities. Accidental launches of nuclear weapons cannot also be totally ruled out, especially, if both nations have scaled up their alert levels. Avoiding scaling up of alert levels would require containment of hostilities to lower levels or, better still, avoidance of conflict. Alerting nuclear weapons must be mutually understood as signifying the red line, for once crossed, political logic could become subservient to the unpredictable, unintelligible and inexorable flow of military activities.

CHAPTER 4

NUCLEAR PAKISTAN

While generally countries have armies, in Pakistan, the Army has a country.

—**Anonymous**

For Pakistan, the lure of nuclear weapons was animated by a deeply-rooted notion in their security imagination—a figment that India's persistent attempt is to seek the demise of Pakistan. Possession of nuclear weapons was perceived as the ultimate bulwark against a bigger and much more powerful India. Terrorism under the nuclear shadow became a foreign policy tool and adopted strategy. Ironically, such an approach has debilitated its own body politic through conversion of a substantial part of its civil society into the jihadi mould and become its primary existential threat. The predicament, albeit partially, is linked to its policy on nuclear weapons.

Any attempt at an analysis of Pakistan's nuclear arsenal can be mainly conjectural due to the clandestineness it maintains, which, coupled with a calibrated exposure is expected to strengthen deterrence. Deliberate ambiguity on its nuclear threshold and capabilities of its delivery systems and warheads are an intentional ploy that Pakistan has adopted so as to boost deterrence. An obfuscation of capabilities finds expression in its projection of missiles as nuclear capable when clearly they are not and also the exaggeration of missile ranges and holdings of missiles and warheads.

While Pakistan's rationale for nuclear weapons is derived from its own perception of the Indian nuclear threat, the Indian nuclear threat, on the other hand, was born of the Chinese threat. The most pervasive internalised narrative of Pakistan's elite has been that India will always attempt to dismember Pakistan. From the 1960s, and especially, after China's first nuclear tests in 1964, Pakistan's leaders have been expressing their apprehensions about India's peaceful nuclear capability being converted into a nuclear weapons capacity. Pakistan's final decision to pursue the building of nuclear weapons is rooted in its deepened sense of insecurity and humiliation experienced after the loss of East Pakistan and the resultant creation of Bangladesh. The internalised historical memory puts the blame for the loss of its eastern half on India. Pakistan thus conveniently blinds itself to its actions that set in motion the course of events for East Pakistanis to separate. Pakistan considers its nuclear arsenal as its crown jewels, which it believes provides ultimate security.

In 1965, Zulfiqar Ali Bhutto, who was an early proponent of nuclear weapons said to the British newspaper *Manchester Guardian*, 'If India makes an atom bomb, then even if we have to feed on grass and leaves—or even if we have to starve—we shall produce an atom bomb as we would be left with no other alternative. The answer to an atom bomb can only be an atom bomb.'[1] When he took over power in 1971, he shaped what has been considered Pakistan's enduring dominant strategic beliefs about nuclear weapons:[2]

- Nuclear weapons provide the only guarantee of Pakistan's

[1]Khan, Feroz Hassan. *Eating Grass, The Making of the Pakistani Bomb*, Foundation Books, 2013, p 7.

[2]Ibid, p 6.

national survival against a perpetually hostile India and unreliable allies.
- Pakistan's nuclear quest is being discriminated against because it is a Muslim country.
- The US, India and Israel may use force to stop Pakistan from acquiring nuclear weapons.

Collectively, these beliefs have continued to animate Pakistan's nuclear weapons reasoning, which it believes, provides the security shield necessary for Pakistan to cope with threats from India. Pakistan has, therefore, singled out nuclear weapons as the means of safeguarding its national security and political integrity. In view of the deep-rooted India–Pakistan animosity harboured by large segments of both societies that is based on a sense of visceral hatred towards each other, manifested in persistent disputes and conflicts between the two countries, the Pakistanis believe that its very existence will continue to be seriously threatened by India. So, by developing nuclear weapons and possessing a minimal number of them, Pakistan believes it can deter India from attacking and further dismembering Pakistan as India did in 1971. Pakistan, however, has not articulated a nuclear doctrine, and only various statements of its leaders over a period provide clues to their thinking on the subject.

Some elements of Pakistan's nuclear policy include:
- To deter India from nuclear and conventional aggression.
- To build nuclear forces to achieve credible minimum deterrence.
- Ensure effective security of nuclear forces against India's strategic assets and retain capacity to retaliate should deterrence fail.
- Build a nuclear triad and concurrently develop robust command and control (C2) and command, control, communications, computers, intelligence, surveillance and reconnaissance (C4ISR).

- Security of nuclear assets from non-state actors and inimical nation states.

Some broad generalisations expressed by Pakistani officials also serve as indicators:[3]

- Nuclear restraint, stability and minimum deterrence constitute the basic elements of nuclear policy.
- Pakistan's nuclear weapons are meant to deter both conventional and nuclear threat from India.
- Pakistan does not subscribe to an NFU policy.
- Nuclear weapons will be used when the security of the nation is jeopardised.

Nuclear Threshold

But the essential Pakistani dilemma still remains in articulating the circumstances under which Pakistan will use its nuclear weapons in the first-strike mode. General Pervez Musharraf declared that nuclear weapons will be used if Pakistan's security is jeopardised.[4] On 26 March 2001, during the India–Pakistan standoff, General Musharraf said that in the prevailing security environment in the region it was imperative for the country to always maintain a 'minimum deterrence'. In an apparent reference to disparity in defence forces vis-à-vis India, he said, 'We are always outnumbered.... We maintain minimum deterrence, which we will always maintain.' Spelling out the three stages of deterrence, General Musharraf emphasised that minimum deterrence, which could be quantified in comparison to the enemy's strength, should be followed by the 'ability to threaten the enemy's such vulnerable targets which go beyond their tolerance threshold'.[5]

[3]Nawaz Sharif's address to National Defence College, Islamabad, quoted in *Pakistan Times*, 20 May 1999.

[4]Z Davis Ed, The India–Pakistan Military Standoff: Crisis and Escalation in South Asia, Macmillian US, 2011, p 74.

[5]Quoted in *Dawn*, Karachi, 27 May 2001. http://www.acronym.org.uk/old/archive/dd/dd56/56ctbt.htm

In 2002, Lt Gen Khalid Kidwai, the head of Pakistan's Strategic Plans Division (SPD), in an interaction with Landau Network-Centro Volta,[6] stated that nuclear weapons will be used only 'if the very existence of Pakistan is at stake'. He defined the circumstances for Pakistan's use of nuclear weapons as:

- India attacks Pakistan and conquers a large part of its territory (space threshold).
- India destroys a large part of its land or air forces (military threshold).
- India proceeds to the economic strangling of Pakistan (economic threshold).[7]
- India pushes Pakistan into political destabilisation or creates a large-scale internal subversion in Pakistan (domestic threshold).

Understandably, there is deliberate ambiguity regarding the interpretations of these parameters and their determination in the use of nuclear weapons. Several Pakistani writers tend to suggest that in case of an Indian attack, the nuclear threshold will be low.[8] A discussion of circumstances for use creates an impression that nuclear weapons are seen more as an extension of firepower to halt an Indian attack.[9] The contemporary Pakistani declarations regarding development of TNWs are a reflection of the 'extension of fire

---

[6]See also report on 'Nuclear Safety, Nuclear Stability and Nuclear Strategy in Pakistan' at https://pugwash.org/2002/01/14/report-on-nuclear-safety-nuclear-stability-and-nuclear-strategy-in-pakistan/

[7]Examples of economic strangling of Pakistan included a naval blockade and the stopping of the waters of the Indus River.

[8]Lodhi, Lt Gen Sardar FS (retd) Pakistan Army, 'Pakistan's Nuclear Doctrine', *Pakistan Defence Journal*, 1999. See also, Agha Shahi, Abdul Sattar and Zulfikar Khan. *Dawn*, 5 October 1999.

[9]Qadir, Brigadier Shaukat. Director, Islamabad Policy Research Institute, quoted by Nair, Brigadier VK. 'National Security Perspectives of India in a Nuclear Backdrop', Seminar Papers, Defence Services Staff College, Wellington, 27-28 February 2001, p 63.

power' paradigm. These discussions never consider the consequences of an Indian reaction to its logical conclusion, which would expose the irrationality of crossing the nuclear threshold under any imaginable circumstances, except as an act of irrationality that countenanced national suicide to achieve political objectives. The first-strike stance is, therefore, fundamentally incredible in circumstances when India also has nuclear weapons. It would virtually be tantamount to a decision that can be described as 'committing suicide due to fear of death'. The situation is exacerbated for Pakistan because of the disparity in the geographic size of the countries and the lack of resources to create a nuclear force that could convincingly achieve a successful first strike against India. For all the bombast Pakistan expends on its first-strike posture, it cannot escape the stranglehold of self-deterrence in a nuclear confrontation because deterrence generally revolves around three parameters—balance of interests, balance of capabilities and balance of resolve. The balance of interests would always weigh against Pakistan as any nuclear use could threaten its existence considering its size differential and the balance of resolve would be in India's favour because Pakistan would have used it first and that would only serve to strengthen India's hand. The balance of capabilities would also always be to India's advantage, considering its indigenous ability in the nuclear arsenal as reflected in the indigenous development of the nuclear-powered ballistic missile submarine.

Pakistan's Utilisation of Nuclear Weapons

The primary focus of Pakistan's arsenal has been on its utility to respond to aggression. There is widespread agreement amongst Pakistan's defence specialists that it is the country's nuclear capability that provides the central basis for strategic deterrence against India. Pakistani analysts often describe nuclear weapons as the 'great equaliser' that offsets India's conventional edge.[10] Prior to 1998, Pakistan bragged that its nuclear programme had deterred India thrice during the period 1984-90. The first in 1984, the second in 1987 during Operation Brass Tacks and the third in 1990.[11] Later, revelations by President Pervez Musharraf prove that in 1999 during the Kargil Conflict, Pakistan's nuclear capability was not yet operationalised and thus exposed the earlier Pakistani bluff.[12]

Pakistan also views the potential for nuclear escalation as a political weapon that provides the means for inviting intervention by outside powers. The Pakistani proclivity to paint the India–Pakistan situation as a nuclear flashpoint has been an obvious ploy whenever tensions increase on the border. Nuclear threats issued at the outset of the Kargil Conflict bear clear evidence of such attempts by Pakistan. On 30 May 1999, soon after India employed air power in the Kargil Conflict, the Pakistani Foreign Secretary Shamshad Ahmad told the media: 'We will not hesitate to use any weapon in our arsenal to defend our territorial integrity.'[13] Consequent to Musharraf's revelations in his autobiography,[14] it is obvious that the threat was not only a bluff but its main intent was to invite intervention by the US.

Some Pakistanis believe that nuclear weapons will not be used and cannot be used for any rational purpose.[15] But without the façade of a first-strike intention, they cannot hope to deter an Indian

---

[10] Quoted from Ghumman, Lt Col (IA), 'Pakistan's Geo-strategic Environment and Military System', *Pakistan Army Journal*, March 1990, p 45, in Giles, Gregory F and Doyle, James E. 'Indian and Pakistani Views on Nuclear Deterrence', *Comparative Strategy*, Vol-15, Issue 2, 1996, p 145.

[11] The Kargil Review Committee Report, op cit, p 194. See also, Ghulam Ishaq Khan's statement quoted in Giles, Gregory F and Doyle, James E. Op cit, p 146.

[12] Musharraf, Pervez. *In the Line of Fire*, Simon & Schuster, London, 2006, p 98.

[13] Baruah, Amit. 'We will not hesitate to use any weapon in our arsenal to defend our territorial integrity', *The Hindu*, 31 May 1999.

[14] Ibid, *In the Line of Fire*, p 98.

[15] 'Tariq Rahman questions about the bomb', *Pakistan Times*, 8 June 1999.

conventional response to their proxy war in Jammu and Kashmir and acts of terrorism in other parts of India. In a psychological sense, the first-strike pretension also serves to buttress their confidence in their own nuclear deterrent capability and is a Pakistani crutch of self-assurance.

One could expect that in any future India–Pakistan conflict, Pakistan would cry for the nuclear wolf as early as possible with the twin political objectives of first, deterring India's conventional application of force, and second, inviting foreign intervention.

The dominant view in Pakistan is that the development of nuclear weapons is linked to the advancement of policy goals in relation to jihad and to achieving Indian concessions on Jammu and Kashmir. The minority focuses on how nuclear weapons currently interfere with Pakistan's economic development.[16] The majority view illuminates how Pakistan seeks to leverage its nuclear weapons in terms of pushing its Jammu and Kashmir agenda. By declaring a low nuclear threshold, it seeks to both deter and shield itself against any Indian conventional response to its proxy war in Jammu and Kashmir. By continuation of the proxy war, it hopes to gain concessions on Jammu and Kashmir. Also, by painting the state as a nuclear flashpoint, it hopes that the US and other Western powers would pressurise India to continue the dialogue that is sought to be influenced by terrorism. Therefore, even though Pakistan's first-strike posture may lack fundamental credibility, it perceives useful political payoffs in its Jammu and Kashmir agenda. It is this challenge that now confronts India's security planners.

Over a period of time, depending on the audience it wants to influence, Pakistan has found several uses for its nuclear arsenal. The important ones are:

- To deter India's conventional military threat.

---

[16] Raghavan, VR. Reporting on Track 2 Dialogue, http://www.thehindu.com/2001/01/25/stories/05252523.htm

- To provide a shield to conduct a proxy war with India in the form of terrorist actions.
- To expedite international intervention in India–Pakistan crisis situations.
- To garner US support for the incumbent government by projecting the possibility of nuclear weapons falling under the control of fundamentalists if they come to power.
- To gain political and economic advantage through nuclear and missile proliferation.
- To provide scope for leadership among Islamic nations by being the sole Islamic nation to possess nuclear weapons.

Path to Development

Pakistan has taken a covert path to develop its nuclear force. The development of its nuclear arsenal has been possible only through clandestine methods because of the nuclear non-proliferation policies of the Western countries. They adopted surreptitious measures to bypass international export controls to acquire enrichment technology and equipment. The extent of Chinese assistance in weapon design and testing as also acquisition of missiles and missile technology is also well known.[17] The 1998 Chagai tests were supposedly based on a weapon design of the fourth Chinese tests conducted in October 1966, which were based on an enriched uranium fission core and had a reported yield of 20-25 kiloton.[18] It is also well known that

[17] The Kargil Review Committee Report, Sage Publications, 2000, pp 183-199. This report covers the assistance received from China for Pakistan's nuclear programme. The Chinese have also admitted to supplying M-11 missiles to Pakistan taking the plea that their range does not violate the MTCR guidelines. See Chinese ambassador to the United States Zhu Qizhen's statement in Di, Hua, 'China's Case: Ballistic Missile Proliferation', in Potter, William C and Jencks, Harlan W, eds, *The International Missile Bazaar: The New Suppliers' Network*, Boulder, Westview Press, 1994, p 169. See also, Jacobs, Gordon and Macarthy, Tim, 'Chinese Missile Sales: Few Changes for the Future', *Jane's Intelligence Review*, December 1992, p 560.

[18] Koch, Andrew and Sidhu, Waheguru Pal Singh. 'Subcontinental Missiles', *Bulletin of the Atomic Scientists*, July/August 1998, p 44.

the US was well aware of the assistance being received by Pakistan but preferred to look the other way during the Cold War, due to the importance of using Pakistan as a conduit to fight the Soviet forces in Afghanistan.[19] Persuasive evidence also exists that North Korea provided missiles and missile technology for the Ghauri missile.[20] Pakistan continues to seek close security ties with China to counterbalance India. Its 1998 nuclear weapons tests and its missile tests over the years are viewed by Pakistan as necessary responses to India's tests and as a means of bolstering its own deterrent.

Musharraf candidly admitted that during the Kargil Conflict, 'I can also say with authority that in 1999, our nuclear capability was not yet operational. Merely exploding a bomb does not mean that you are operationally capable of deploying nuclear force in the field and delivering a bomb across the border over a selected target. Any talk of preparing for nuclear strikes is preposterous.'[21] This statement reflects the reality that exploding a nuclear weapon at a test site is only a major milestone on the long road to achieving a robust operational capability. Operational capability requires national security policy in the form of a nuclear doctrine that will provide the guidance for achieving an envisaged capability. It involves the creation of organisational structures for command and control and extensive executive elements that implement decisions and ensure survival. Such requirements take several decades to fructify and also warrant constant re-evaluation in the face of emerging threats and technological progress in defensive and offensive capability. Limited availability of economic and technological resources is also

---

[19] Spector, Leonard S and Smith, Jacqueline R. *Nuclear Ambitions: The Spread of Nuclear Weapons*, Westview Press, Boulder, 1990, pp 95-96.

[20] Chandrashekhar, S. 'The Origins and Antecedents of the Ghauri Missile: An Assessment', *Current Science*, Vol-76, No. 3, 10 February 1999, pp 280-85. The technical arguments are persuasive. See also 'Pakistan tests Chinese/North Korean-based missiles', *Jane's Intelligence Review*, 5/99, p 3.

[21] Ibid, *In the Line of Fire*, pp 97-98.

a major challenge. Pakistan has made considerable progress in bolstering its nuclear weapons capability that includes delivery systems, warhead design, fissile materials availability, command and control and survivability of nuclear wherewithal. It would, however, be reasonable to assume that it still has much distance to travel on the road towards robust operational capability.

Delivery Systems: Aircraft

Pakistani sources claim that by mid-1995 it had cold-tested a nuclear device based on a Chinese design deliverable by fighter aircraft (F-16s) and the Mirage-V later supplemented this capability. This was followed by the development of various bombing techniques. The manoeuvres included 'conventional free fall', 'loft bombing', 'toss bombing' and 'low-level' attack techniques.[22] The F-16 has a range of 630 km in a 'hi-lo-lo-hi' profile. These aircraft could effectively cover the states of Gujarat, Rajasthan, Delhi, up to Agra in Uttar Pradesh and Uttarakhand. Some sources have indicated that Pakistan did not have the means to arm the nuclear weapons while in flight, since the software of the F-16 aircraft was not compatible with nuclear or for that matter any other ordinance placed on the underwing pylons. Over time, these problems may have been resolved and possibly one squadron of F-16s may be available for nuclear role. But with limited range and vulnerability of aircraft, Pakistan has attempted to compensate by relentlessly pursued ballistic and cruise missile capability.

Ballistic Missiles

The main thrust of development in Pakistan's nuclear delivery capability has been land-based missiles. Pakistan's arrival as a nuclear-capable nation was announced in February 1989 with the flight tests of two Hatf missiles claiming a payload of 500 kg, range of 500 km

[22] Ibid, *Eating Grass: Making of the Pakistani Bomb*, pp 186-87.

and being extremely accurate systems. The bluff was evident when the Hatf series was shelved for more than a decade until February 2000, when two missiles Hatf-IA and Hatf-II (Abdali) were tested claiming a payload of 500 kg and ranges of 100 km and 180 km, respectively. Though the Abdali was tested in 2002 and inducted subsequently, its accuracy has been doubted and contrary to claims, it is unlikely that it can carry nuclear warheads, which fits into the known pattern of Pakistan declaring all its missiles as nuclear-capable to deepen ambiguity and strengthen deterrence.[23]

In the early 1990s, through transfer of technology, Pakistan bought the M-9 and M-11 (which the Chinese refer to as the Dongfeng, DF-15 and DF-11). The Chinese M series of tactical short-range ballistic missiles were developed in the early 1980s. The three known versions are the M-9, M-11 and M-18; the M designations are used for the export versions. The M-11 missile was first revealed at a Chinese exhibition in 1988, and was displayed with a transporter erector vehicle similar to that used with the M-9 (DF-15). The M-11 is a single-stage solid-fuel missile with terminal guidance, a range of 300 km and a payload capability of about 500 kg[24] and the M-9 missile is also solid fuelled with a range of 600-700 km. According to the Kargil Review Committee Report, by 1991, the Chinese had begun directly transferring M-11 missiles to Pakistan and by 1984, eighty-four of them were transferred. These were stored at the Central Ordinance Depot at Sargodha and were assigned to the 155 Composite Regiment of Second Army Artillery Division, Attock.[25] Some reports indicated that some missiles were also at Samungli in Quetta. Chinese experts had also reportedly imparted training to the Pakistani unit.[26]

---

[23] Khan, Feroz Hassan. op cit, pp 237-38.

[24] www.cdi.org/issues/nukef&f/database/panukes.html

[25] The Kargil Review Committee, op cit, p 192.

[26] www.cdi.org/issues/nukef&f/database/panukes.html

Pakistan claims that both the M-9 and M-11 missiles were tipped with high-explosive warheads and their scientists redesigned the missile to make it nuclear capable. This required extensive research and consequent modifications. The Ghaznavi (Hatf-III), based on the M-11, and the Shaheen-I and IA (Hatf-IV), based on the M-9, are the end products and are supposedly capable of carrying Pakistan's second-generation nuclear warheads.

Shaheen-I

The Shaheen series had two configurations. Shaheen-I with an overall length of 11.53 m has a range of 700 km, payload of 1,000 kg and has been tested the maximum number of times indicating that this version has been introduced into service. Shaheen-IA was tested at least four times till 2016 but has probably not yet been introduced into service.

Ghaznavi

Derived from the M-11, Ghaznavi could carry a payload of 500-800 kg with a range of 280-320 km. It was first tested in 2002 and its final version probably became fully operational after further tests in 2007.[27] The Ghaznavi and Shaheen-I were produced in the Chinese-built missile factory at Fatehjung and laid the foundation for producing future systems especially based on solid propellants.[28]

Ghauri

Pakistan's search for longer-range missiles, with greater payload to cover the entire Indian peninsula, and its need to diversify sources led it to seek North Korean assistance. China was unwilling to supply longer-range missiles due to its commitment made under the Missile Technology Control Regime (MTCR) to not transfer missiles with

---

[27] Chandrashekhar, S and Nagappa, Rajaram. 'Pakistan's Ballistic and Cruise Missiles', International Strategic and Security Studies Programme, National Institute of Advanced Studies, Bangalore, February 2013. See also Khan, Feroz Hassan, op cit, pp 239-40.

[28] Ibid, *Eating Grass: The Making of the Pakistani Bomb*, p 242.

ranges longer than 300 km. Starting in 1994, Pakistan purchased the liquid-fuelled North Korean Nodong missile that was renamed Hatf-V (Ghauri). The unassembled missiles and transporter erector vehicles were delivered in 1997. Assembly facilities were established at Kahuta. North Korea, therefore, became the technology supplier for Pakistan's liquid fuel missiles. Pakistan in April 1999 claimed a range of 2,000-3,000 km, which was an obvious exaggeration. The Pakistani Government has been quoted as saying that the Ghauri 'can be tipped with any type of warhead'.

Based on comparative technical parameters, the missile was almost certainly the North Korean Nodong missile, which is a derivative of the Russian Scud-B by clustering.[29] The guidance systems for the Ghauri are likely to have originated in China.[30] Indian technical analysis indicates that the Ghauri is a single-stage liquid-fuelled missile capable of achieving a range of 950-1,020 km with a load of 700 kg–1 ton,[31] though some analysts say that the maximum range is only 700 km.[32] The Pakistanis claim a range of 1,500 km with 700 kg payload. Pakistani scientist AQ Khan, however, claimed that the Ghauri could reach all Indian cities[33] which was a gross exaggeration. With a maximum range of 1,000 km, it could cover parts of central India including Hyderabad and Lucknow. The Ghauri has been tested five times after deployment. During the test in November 2012 it exploded during the terminal phase. But a

---

[29] Chandrashekhar, S. 'The Origins and Antecedents of the Ghauri missile: An assessment', *Current Science,* Vol-76, No. 3, 10 February 1999, pp 280-285. See also Wright, David C. 'An Analysis of the Pakistani Ghauri Missile Tests of 6 April 1998', Union of Concerned Scientists and Security Studies Program, MIT, 6 May 1998. http://www.fas.org/news/pak/1998/05/index/html. Accessed 24 March 2001.

[30] http://www.b14643.de/Spacerockets/Specials/Pakistan-China/index.htm

[31] Chandrasekhar, S. 'The Origins and Antecedents of the Ghauri Missile: An Assessment', *Current Science,* Vol-76, No. 3, 10 February 1999.

[32] See note 29, David Albright's 'Analysis of the Tests'.

[33] http://www.thehindu.com/2000/09/14/stories/03140007.htm

successful test in April 2015, indicated that the problem had been rectified.

Shaheen-II

While the Ghaznavi and Shaheen are by Pakistan's admission derived from the Chinese missiles, the longer range (2,000 km) Shaheen-II (Hatf-VI) is claimed to be indigenous. This claim is contested by other sources that maintain that the Shaheen-II is derived from the Chinese M-18.[34] It was first displayed in October 2003 and first flight tested in March 2004 and claimed to have covered a range of 1,880 km while the potential range claimed was 2,500 km. This in 2004 was an obvious exaggeration. Indian analysts, however, estimated the potential range to be 1,300 km with a 1,000 kg payload. The Shaheen-II was tested six times up to 2008. After a gap of six years it was again tested in 2014. The claimed range was scaled down to 1500 km.[35] It is a two-stage solid-fuel missile and road mobility is based on a 12-wheeled transporter erector launcher. The Shaheen-II would require additional tests before it is operationally deployed.

Shaheen-III

The Shaheen-III is a two stage missile that was first tested in November 2015 and again in December 2015 and on both occasions it achieved a range of 2,750 km which is the longest for any Pakistani missile. It has a diameter of approximately 19 m and is similar in appearance to the Shaheen-II though slightly bigger. The Shaheen-III can carry a warhead of 1,000 kg and target the entire Indian landmass except possibly the extreme parts of the Andaman and Nicobar Island territories. Some more testing would be probably required before being operationally deployed.[36]

---

[34] 'Strategic Weapons System', *IHS Jane's Sentinel Security Assessment-South Asia*, 2014. http://www.b14643.de/Spacerockets/Specials/Pakistan-China/index.htm

[35] ISPR Press Release No 248/2014 dated 13 November 2014.

[36] ISPR Press Release No PR-61/2015-ISPR dated 9 March 2015 and PR-378/2015-ISPR dated 11 December 2015. www.ispr.gov.pk

Ababeel

On 24 January 2017, Pakistan claimed to have achieved MIRV capability with the maiden launch of a new missile called the Ababeel. The ISPR release claimed that 'the Ababeel is capable of carrying nuclear warheads and has the capability to engage multiple targets with high precision, thus defeating the enemy's hostile radars. The development of the Ababeel weapon system is aimed at ensuring the survivability of Pakistan's ballistic missiles in the growing regional BMD environment. This will further reinforce deterrence'.[37]

The claimed range of the Ababeel is 2,200 km. Images indicate that it is a three-stage missile with the first two stages similar in dimension to the Shaheen-3 and a bulbous third stage on top. The unusually shaped third stage is possibly to accommodate the multiple warheads.

Pakistan's claims of achieving MIRV capability with the very first launch of such a missile is evidently an exaggeration. Notably, there is a difference between delivering multiple warheads over a given target and independently targeting widely separated targets with the same missile. Experts have questioned the MIRV capability of the Ababeel including the claimed range of 2200 km. Above all, the real question is the nuclear capability of the missile as it is contingent upon successful miniaturisation of warheads, for which there is no credible evidence. Considering the requirement of several more tests, the Ababeel's operational deployment could take place earliest by 2021-22.

Notwithstanding the uncertainties associated with the Ababeel, the quest to acquire MIRV capability indicates Pakistan's apprehensions about the Indian BMD programme and the credibility of its deterrent. It also signals a new form of missile race in the subcontinent.

[37] Press Release No 34/2017 dated 24 January 2017. www.ispr.gov.pk

Table 4.1: Pakistani Ballistic Missiles-2017

Missile	Type	Fuel	Range	Payload	First Test	Total Tests Till Date	Entry Into Service	Approx No	Remarks (External Assistance)
Nasr	SRBM	Solid	60 km	300 kg	2011	6	?	?	Under Development
Ghaznavi	SRBM	Solid	300 km	700 kg	2002	12	2004	24-36	Chinese M-11
Ghauri	MRBM	Liquid	1250-1500 km	700 kg	1998	10	2003	12-24	North Korean Nodong
Shaheen-1	MRBM	Solid	750 km	700 kg	1999	12	2004	12-24	Chinese M-9
Shaheen-1A	MRBM	Solid	900 km	700 kg	2012	04	?	?	-
Shaheen-II	MRBM	Solid	1500-2000 km	1000 kg	2004	7	2008	8-12	Chinese M-18
Shaheen-III	MRBM	Solid	2750 km	1000 kg	2015	2	-	-	Under Development
Ababeel	MRBM	Solid	2200 km	1000 kg	2017	1	-	-	Under Development

Compiled from Centre for Strategic and Internatinal Studies, https://missilethreat.csis.org/country/pakistan/

Ballistic Nuclear Missile Inventory

Indian technical estimates of Pakistan's ballistic missiles inventory indicate that only the Ghaznavi, Shaheen-I, Shaheen-II and Ghauri could be capable of carrying nuclear missiles based on Pakistan's assessed capacity to miniaturise. They also estimate that by 2011, Pakistan's inventory of ballistic missiles as Ghaznavi, thirty-five; Shaheen-I, fourteen; Shaheen-II, six; Ghauri, eleven, to be a total of sixty-six.[38] This estimate is based on Pakistan's assessed capacity for propellant grain production of twelve propellant grains per year. The Ghaznavi numbers assume the transfer of thirty-four missiles from China. However, some estimates indicate the transfer of eighty missiles. Moreover, as Shaheen-II is unlikely to be operational, the estimate could be in the range of 60-106.

Pakistan claims of possession of other nuclear-capable missiles—Abdali, Shaheen-IA, Babur, Ra'ad and Nasr are highly questionable due to issues of miniaturisation of warheads and fissile material availability.

Cruise Missiles

The development of cruise missiles was a natural transit point in Pakistan's quest for the diversification of its missile armoury and the increase in the effectiveness of its delivery systems. The terrain-hugging cruise missiles can easily evade radar detection and are also small enough to be mobile and easy to conceal, thereby increasing their survivability. Pakistan's cruise missile development has been credited to the reverse engineering skills of Pakistani and Chinese scientists and became feasible because Pakistan supposedly gained access to American Tomahawk missiles when they were fired from US ships in the Arabian Sea at terrorist camps in Afghanistan in

---

[38] Chandrasekhar, S and Nagappa, Rajaram. 'Pakistan: Ballistic and Cruise Missiles', International Studies and Security Studies Programme, National Institute of Advanced Studies, Bangalore, February 2013.

August 1998. Some unknown number of missiles malfunctioned and fell in Balochistan and Pakistan recovered some missiles that were almost intact. These missiles may also have been shared with the Chinese who derived the DH-10 from them.

Hatf-VII (Babur)

The first test of Hatf-VII (Babur) took place in 2005 and was followed by more in 2006, 2007, 2009, 2011 and 2012. After the 2012 launch, official Pakistani agencies described the missile as having stealth features, being equipped with the modern cruise missile technology of terrain contour matching (Tercom), digital scene matching and area correlation (DISMAC), and the capacity to carry both conventional and nuclear warheads up to a range of 700 km.[39]

Analyses of the Babur by NIAS, Bangalore, indicate that it could achieve a range of 700 km. Though Pakistan claims that it will develop a submarine-launched version of the Babur, the diameter of the Babur is too large to be accommodated in the torpedo tubes of the Agosta submarines and hence, will be suitable only as ground-launched cruise missile (GLCM). The possibility of nuclear tipping the missile is discounted. They doubt Pakistan's technical capability to miniaturise the warhead with its associated electronics and trigger system into a compact mass of 450-500 kg. This is based on an earlier study by NIAS on the Ghaznavi warhead sizing which had estimated the warhead mass to be around one ton. They, therefore, doubt the feasibility of fielding a smaller plutonium-based weapon in the absence of a qualifying test.[40]

---

[39] http://www.ispr.gov.pk/front/main.asp?o=t-press_release&id=2150. Earlier, after tests on 10 February 2011 the range stated was 600 km. It was stated as 700 km after tests on 28 October 2011 and 17 September 2012.

[40] Chandrashekhar, S and Nagappa, Rajaram. 'Pakistan's Ballistic and Cruise Missiles', International Strategic and Security Studies Programme, National Institute of Advanced Studies, Bangalore, February 2013.

Hatf-VIII (Ra'ad)

The Ra'ad is an air-launched cruise missile (ALCM), claimed as an indigenous product,[41] and was initially tested on 27 August 2007 from a Mirage-III aircraft with a claimed range of 350 km and supposed nuclear capability. It has been tested seven times with the latest test in January 2016. Indian scientists at NIAS, Bangalore, doubt the veracity of the Ra'ad being capable of carrying a nuclear warhead based on their evaluation of Pakistan's nuclear tests, the payload and mass envelope.[42]

Hatf-IX (Nasr)

The Hatf-IX (Nasr) is the latest addition to Pakistan's delivery systems. It was first flight tested on 19 April 2011. Further tests were carried out on 29 May 2012, 11 February 2013, 5 November 2013, 26 September 2015 and 4 July 2017. Identical official press releases described the missile as 'short-range surface-to-surface ballistic missile; the missile has been developed to add deterrence value to Pakistan's Strategic Weapons Development Program at shorter ranges and carries a nuclear warhead of appropriate yield with high accuracy, shoot and scoot attributes'.[43] The Nasr is essentially a rocket, launched from a four-tube mobile transporter erector launcher (TEL). It has a fairly low trajectory unlike a ballistic missile. The Nasr triggered a debate about feasibility in terms of warhead miniaturisation, its utility and its impact on deterrence stability.

Nasr has a diameter of 350 mm, a conical portion which is 660 mm and a cylindrical section which is 940 mm. Indian technical analysis of Pakistan's capability to miniaturise the warhead to fit the

---

[41] Ibid, *Eating Grass: The Making of the Pakistani Bomb*, p 248.

[42] Chandrashekhar, S and Nagappa, Rajaram. 'Pakistan's Ballistic and Cruise Missiles', International Strategic and Security Studies Programme, National Institute of Advanced Studies, Bangalore, February 2013.

[43] See Inter Services Public Relations press release, 19 April 2011, http://www.ispr.gov.pk/front/main.asp?o=t-press_release&id=1721

Nasr indicates that while it is theoretically possible to do so, it can only be done by a plutonium-based linear implosion device. Such a device requires nearly double the quantity of plutonium. A similar design required extensive experimentation in the US and it is unlikely that Pakistan could achieve such a degree of miniaturisation. Pakistan had also only tested highly enriched uranium (HEU)-based devices in 1998 and none of those devices fits the Nasr. So, even if Pakistan has managed to miniaturise a plutonium-based warhead for Nasr, it would be an untested weapon with questionable reliability.[44]

Pakistan's limitations of fissile material availability also create doubts regarding the possibility of utilising its limited uranium output for plutonium reprocessing. According to calculations of the NIAS study of 2013, Pakistan's plutonium inventory will be adequate for 12 warheads for Nasr, and even if its total uranium output is diverted for plutonium processing, Pakistan will be able to add only 5-6 warheads per year. Pakistan will, therefore, take several years to build a reasonable-sized arsenal for use as battlefield weapons. This will also impede any addition to the rest of its arsenal. Nasr carries an untested device that is supposedly meant for battlefield use and this is coupled with the lack of fissile material availability. There are also serious doubts about Pakistan's claims regarding the Nasr, due to its damage potential and the need to have large quantities for effectiveness on mechanised targets.

In a study conducted by AH Nayyar and Zia Mian, it has been technically proved that Nasr's damage potential is rather limited,[45]

---

[44] Nagappa, Rajaram; Vishwanathan, Arun and Malhotra, Aditi. 'Hatf-IX/Nasr-Pakistan's Nuclear Weapon: Implications for India–Pakistan Deterrence', International Strategic and Security Studies Programme, National Institute of Advanced Studies, Bangalore, July 2013, pp 21-22.

[45] Nayyar, AH, and Mian, Zia. 'The Limited Military Utility of Pakistan's Battlefield Use of Nuclear Weapons in Response to Large-scale Indian Conventional Attack', Pakistan Security Research Unit, Brief Number 61, Department of Peace Studies, University of Bradford, UK, November 2010.

and it was estimated that with a weapon yield of 15 kt, eight weapons would be required to incapacitate fifty-five tanks. Considering India's offensive potential, Pakistan would require significant quantities of battlefield weapons; such quantities are well beyond Pakistan's resources.

Rationale for Battlefield Nuclear Weapons

Even if it is a bluff, the Nasr's claim can be verified, by Pakistan's rationale for battlefield nuclear weapons. The rationale for weapons like the Nasr is based on Pakistan's need to bridge the tactical and operational gap supposedly created by India's Cold Start doctrine. In March 2013, the Director General of Strategic Plans Division, Lt Gen Khalid Ahmed Kidwai, argued that Pakistan's TNW programme was a result of the widening gap created on account of India's massive weapons acquisition, huge defence budget, Cold Start doctrine, ballistic missile defence and the Indo-US nuclear deal. As a result, Pakistan was trying to 'plug the tactical and operational gaps' caused by India's nuclear and conventional strategy.[46] The Cold Start doctrine is assumed to be India's answer to the problems experienced during the mobilisation of the Indian armed forces during Operation Parakram in 2002. The Operation was launched consequent to the attack on the Indian Parliament. Mobilisation, especially, of offensive ground forces took several weeks, giving Pakistan time to prepare and also facilitated international intervention. The Cold Start doctrine required the acquisition of the capability to launch limited ground offensives based on defensive formations that were located *a priori* in the vicinity of the India–Pakistan border. Such a capacity would limit itself to shallow thrusts to deny Pakistan any justification to escalate to the nuclear level and inflict significant damage to Pakistan before intercession by the international community. Acquiring such

[46] https://www.iiss.org/en/events/events/archive/2013-5126/december-c771/defence-doctrines-workshop-0298

capacity would involve the relocation of considerable offensive assets now located in central India. In any case, the Cold Start idea was initially a uniservice product that was voiced by the Army and did not have the support of the air force and the navy. Although the doctrine was militarily and politically disowned in 2010,[47] it was, resurrected by the Indian Army Chief, General Bipin Rawat, in January 2017.[48]

Pakistan has continued to internalise Cold Start and finds it convenient to quote its existence to magnify the Indian threat. The rationale for the Nasr, based on the Cold Start threat, is questionable because the Cold Start involves ground actions that are shallow thrusts much below Pakistan's avowed thresholds. Instead of using nuclear weapons as a last resort, battlefield use could mean targeting spearheads or logistical bases of the Indian offensive.

India, on the other hand, has avowed that any use of nuclear weapons will invite nuclear retaliation that will be massive and designed to inflict unacceptable damage. India does not distinguish between tactical and strategic nuclear weapons. It considers nuclear use to be the sole criteria. This is based on the notion that there is nothing intrinsic in a weapon to make it tactical or strategic. Even a bullet fired from a pistol that kills an important world leader could have strategic effect. The effect is not inherent in the weapon but in how it is used and what an aftermath it creates. The very use of nuclear weapons, even if they are only 1-2 kilotons, will have a strategic effect when used on any targets even in the battlefield. TNW as a separate type of weapon is not recognised by India.

Sea-based Capability

In 2012, Pakistan announced that it had established the Headquarters Naval Strategic Forces Command and projected it as the initial

---

[47] Subramanian, NV. 'India Denies "Cold Start" Plan', *The Diplomat*, Tokyo, 11 September 2010. https://thediplomat.com/2010/09/india-denies-cold-start-plan/

[48] Interview with Sandeep Unnithan, *India Today*, 4 January 2017.

step to develop and deploy the sea leg of the triad. The official announcement described it as becoming the 'custodian of second-strike capability' to 'strengthen Pakistan's policy of credible minimum deterrence and ensure regional stability'. The Babur cruise missile is supposedly being modified to be fired from a sea-based platform. Possibly, the Babur could be retro-fitted on a naval ship. But it is unlikely to be fitted on any of the existing submarines or the eight Yuan class diesel electric submarines contracted from China in 2016 with a delivery schedule of four by 2023 and the remaining four in 2028. A ballistic missile nuclear powered submarine is presently and in the foreseeable future unlikely to be a component of its nuclear weapons wherewithal—bluff and bluster notwithstanding.

Missiles—Creating Ambiguity

Diverse public pronouncements by various Pakistani officials have tended to obscure rather than clarify the present status of Pakistan's missile programmes. The confusion is greatly magnified by the diversity of nomenclature, as the number of missile names, evidently greatly exceeds that of the actual missile types. It is obvious that the creation of confusion is deliberate and intended to veil the truth regarding Pakistan's actual capabilities. Pakistan's missile tests have mostly been an immediate reaction to India's missile tests. The various statements emanating from Pakistan are indicative of their desire to project an ability to cover entire India with the range capability of its missiles. It would require a 2,000-2,500 km range missile which could be realised when the Shaheen-III and the Ababeel are operationally deployed. Presently, Pakistan probably has three operational nuclear missile systems—Ghaznavi, Shaheen-I and Ghauri. The Pakistani missile bluff is obvious and an ingenuous attempt to mislead. Leaked stories and tangential statements cannot substitute for a capability that needs to be demonstrated through a soaring missile whose technical parameters can be subjected to examination through satellite tracking. However, for planning purposes, India would be well advised to assume that sooner than later Pakistan will be able to operationalise missiles that will cover the entire country.

The ability to launch missiles does not in itself automatically mean a technological capability to deliver nuclear warheads. Successful nuclear delivery is contingent on miniaturisation of warheads and mastery of re-entry technology. In the long run, however, Pakistan's nuclear capability will be hostage to its relations with its foreign suppliers. One cannot see this strategic weakness of Pakistan being reduced given its economic situation and state of indigenous technological capability. China's assistance in its missile development, warhead design capability and command and control including safety issues can be expected to continue. Pakistan will, in all probability, constantly attempt to bolster the deterrent power of its nuclear arsenal through falsifying its nuclear weapons inventory and delivery capability. Nuclear bombast from its officials like hitherto fore could be expected to continue.

Warheads

China has in all probability provided Pakistan with the blueprints to the Uranium implosion device (CHIC-4) that it detonated on 27 October 1966. It is also suspected that China tested a CHIC-4-derived Pakistani bomb on 26 May 1990 at its Lop Nor site with a yield of 10-12 kiloton range. This range matches Pakistan's 1998 nuclear tests which had yields between 5-12 kiloton. Thereafter, it is reasonable to assume that weapons improvements have been underway to increase the efficiency of the weapons and miniaturisation to arm its cruise and ballistic missiles. But its ability to do so based on plutonium without testing is questionable and improbable.

FISSILE MATERIAL

Estimates of Pakistan's nuclear fissile material capacity vary widely. The International Panel on Fissile Materials estimated that by 2013 Pakistan would have an accumulated HEU stockpile of 3,000 kg and plutonium stockpile of 150 kg. They estimated that Pakistan's reactors could each produce 6-12 kg of plutonium per year or equivalent to one-three nuclear weapons per year depending on the weapon design.[49] This could increase to 24-48 kg whenever Khushab-III and IV are operational. The enriched HEU capability is estimated to correspond to 10-15 warheads per year. The international panel estimated that by early 2013 Pakistan might have a stockpile of material sufficient for more than 200 weapons, even as it produced material for about twelve to twenty-one weapons per year. It has the capacity to increase to fourteen to twenty-seven weapons per year when all four reactors are operational. However, with four reactors then in operation, by 2016 (Khushab I, II and III, IV), they could have an accumulated stockpile estimated as about 280 kg of plutonium and 3.4 tons of HEU.[50]

The requirement of fissile material for pure fission nuclear weapons depends on technical capability and the yield desired. According to Cochran and Paine, the fissile material requirement for pure fission weapons is as given in Table 2.[51]

---

[49] http://fissilematerials.org/countries/pakistan.html

[50] Pakistan-International Panel on Fissile Materials, www.fissilematerials.org

[51] Cochran, Thomas B and Paine, Christopher E. 'The Amount of Plutonium and High-Enriched Uranium Needed for Pure Fission Nuclear Weapons', National Resources Defense Council Inc., Washington DC, 13 April 1995. http://ambienteparco.it/wp-content/uploads/2017/08/fissionweapons.pdf

Table 4.2: Approximate Fissile Material Requirements for Pure Fission Nuclear Weapons

Weapon-grade plutonium (kg) Technical-capability level				HEU (kg) Technical-capability level		
Yield (kt)	Low	Medium	High	Low	Medium	High
1	3	1.5	1	8	4	2.5
5	4	2.5	1.5	11	6	3.5
10	5	3	2	13	7	4
20	6	3.5	3	16	9	5

Another study by Zia Mian, AH Nayyar and R Rajaraman estimates that by 2020 Pakistan could have accumulated approximately 450 kg of plutonium from the Khushab reactors and 2,500-6,000 kg of HEU (90 per cent enriched). These stocks, they calculate, would be sufficient for approximately 100-240 simple fission weapons based on HEU and for ninety plutonium weapons. However, Pakistan could increase this capacity if it either increases the rate of uranium mining or improves its warhead designs.[52]

According to another study by Tamara Patton, the estimate of plutonium as of December 2012 from Khushab-I and -II was 143 kg and 30 kg, respectively.[53] Khushab-I has been active since 1999, Khushab-II since 2008 and Khushab-III since May 2012. Hence, Khushab-III would have produced another 6-12 kg. Construction of Khushab-IV was completed by 2015. Rajaram Nagappa of NIAS, Bangalore, estimates that assuming 8 kg of plutonium per weapon, this would translate to 22 warheads in stock and an annual

[52] Mian, Zia; Nayyar, AH and Rajaraman, R. 'Exploring Uranuim Resources Constraints on Fissile Material Production in Pakistan', *Science and Global Security*, Vol-17, Issue 2-3, 2009, pp 77-108.

[53] Patton, Tamara. 'Combining Satellite Imagery and 3D Drawing Tools for Non-proliferation Analysis: A Case Study of Pakistan's Khushab Plutonium Production Reactors', *Science and Global Security*, Vol-20, Issue 2-3, 2012, pp 117-214.

production of five-six weapons per year and this includes nineteen kg/per year estimated plutonium production from Khushab-III also taken into account.[54]

By 2015, estimated accumulated stockpile was about 0.19 tons of plutonium and 3.1 tons of HEU. It could add 0.04 tons of weapons grade plutonium to its inventory annually with Khushab I, II, III, IV being operational. For uranium enrichment, Pakistan has a centrifuge plant at Kahuta with some reports indicating that it also has another one at Gadwal.[55] The *Bulletin of the Atomic Scientists* in 2016 estimated that Pakistan could be adding about 10 warheads to its arsenal every year. Khalid Kidwai, the erstwhile head of the Pakistan's SPG, indicated that Pakistan had increased its initially planned quantity of warheads to accommodate what he described as a switch from credible minimum deterrence to full spectrum deterrence. This is a switch that was supposedly undertaken to deal with India's Cold Start doctrine. He further indicated that 90-95 per cent of warhead production targets planned fifteen years ago had been met.[56] Pakistan being the singular hold-out in Fissile Material Cut-off Treaty (FMCT) negotiations at Geneva is indicative that it requires more time to accumulate fissile material for the planned arsenal.

A large variance in calculating Pakistan's fissile material and quantity of warheads is inevitable as it is based on assumptions about size of reactors, reprocessing capacity, uranium ore availability, uranium enrichment capacity, warhead design proficiency, production capacity, operational nuclear delivery vehicles, fissile material reserves and the

---

[54] Rajaram Nagappa. 'Technical Aspects of HATF-9/Nasr Missile' in Kanwal, Gurmeet and Chansoria Monica, eds. *Pakistan's Tactical Nuclear Weapons Conflict Redux*, Knowledge World in association with Centre for Land Warfare Studies, New Delhi, 2014, p 166.

[55] https://www.armscontrol.org/files/Threat_Assessment_Brief_Nuclear_Weapons_Geometry_of_Asia.pdf

[56] Kristensen, Hans M and Norris, Robert S. Pakistan Nuclear Forces, 2016. *Bulletin of the Atomic Scientists*, 72:6, 368-76, DOI:10.1080/00963402.2016.

statements of government officials. Moreover, bluff and bluster has been the leitmotif of Pakistan's nuclear capability projection.

Command and Control

Pakistan's nuclear decision process initially remained closely guarded and opaque. Its main characteristic is the pervasive influence of the Pakistani military on nuclear decision-making. The initial plans for setting up a command and control organisation were announced by General Pervez Musharraf in April 1999. Eventually, after the military takeover, Pakistan announced the formation of a National Command Authority (NCA) on 2 February 2000 to control the policy on nuclear weapons with the President as the chairman and the Prime Minister the deputy chairman. According to the official announcement, the NCA is responsible for policy formulation and will exercise employment and development control over all strategic nuclear forces and strategic organisations. It consists of an employment control and a development control committee. The Strategic Plans Division acts as its secretariat. In December 2007, Musharraf passed the National Command Authority Ordinance, 2007, that formalised the NCA. The ordinance includes the issues of personnel reliability and proliferation of technical know-how. In 2009, Asif Ali Zardari as President revised the chairmanship and the Prime Minister was appointed chairman. Thereafter, there has been no deputy chairman of the NCA.

The Employment Control Committee (ECC) is chaired by the head of the government and includes the ministers of foreign affairs (deputy chairman), defence, interior, finance, chairman of the joint chiefs of staff committee (CJCSC), service chiefs, director-general of strategic plans division (secretary) with technical Advisors and others, as required by the chairman. The ECC is responsible for threat evaluation, setting of strategic goals and the authorisation for use of nuclear weapons.

The development control committee is also chaired by the head of government and includes CJCSC (deputy chairman), services chiefs, director general of Strategic Plans Division and heads of concerned strategic organisations. This committee controls the development of strategic assets. It also maintains oversight on the financial, administrative and technical workings of all nuclear organisations.

The Strategic Plans Division, headed by a senior army officer, was established in the Joint Services Headquarters under the CJCSC to act as the secretariat for the NCA and perform functions relating to planning, coordination and establishment of a reliable command, control, communication, computers and intelligence network.[57]

Operational nuclear assets are under the Strategic Forces Commands (SFC) that have been established in each service. Though public announcements of the creation of these commands have been made, the degree of progress in their creation is not known. The respective services are responsible for the training, technical control and administration of their strategic force. Operational control is exercised by the NCA and Lt Gen Khalid Kidwai emphatically reiterated at an International Institute for Strategic Studies (IISS) workshop that Pakistan will not pre-delegate command and control of TNW to field commanders.[58]

The protocol for weapon launch follows the 'two-man rule' that entails two nominated officers accessing and authenticating launch codes. Warheads are also de-mated from delivery systems. Mating is controlled by the NCA. Pakistan has also reportedly developed its own version of permissive action links (PALs) to prevent unauthorised use of nuclear weapons. A jihadi group that seizes a nuclear weapon will find it difficult to use the weapon, as attempts to arm the weapon are likely to trigger safety mechanisms that will disable it. The greater

---

[57] 'National Command Authority Established', Associated Press of Pakistan, 3 February 2000. See also, *The Times of India*, New Delhi, 4 February 2000.

[58] Conversation with a participant, who preferred non-attribution, in the workshop.

danger in Pakistan is the probability of its personnel in the security apparatus being compromised. The possibility of terrorist groups gaining access to toxic fissile materials, such as radioactive waste plutonium from spent fuel, cannot be discounted. A dirty bomb based on a radioactive dispersal device can be used by terrorist groups.

Military Influence

The stranglehold of the Pakistani military on strategic decision-making is well known and adds to its unstable image.[59] In fact, it is an image that suits their intentions. The military control of Pakistan's nuclear arsenal has remained even when a civilian government has been in power. They will seek to exploit this control in a crisis by issuing threats that they hope will carry more credibility because it is controlled by the military. It is, therefore, a convenient ploy to flash the nuclear card and invite international intervention. Recent declarations regarding battlefield nuclear weapons are another attempt to speed up international intervention and stymie India's concept of a limited war.

Pakistan's Credibility Problem

Credibility is the biggest problem with a nation that threatens first use since it not only has to cater for reaction but also face the international opprobrium of breaking the tradition of nuclear taboo that has been maintained for the last six decades. The Pakistani idea is to create an element of doubt that it hopes will strengthen deterrence against the application of India's conventional military power. But deterrence depends on credibility. Credibility in this instance is weakened by Pakistan deciding to start a process (early use of nuclear weapons) that has all the chances of leading to its own destruction,

---

[59] Krepon, Michael. 'Pakistan's Nuclear Strategy and Deterrence Capability', The Stimson Center, Washington DC, 10 December 2012. http://indianstrategicknowledgeonline.com/web/Krepon_-_Pakistan_ Nuclear_Strategy_and_ Deterrence_Stability.pdf

apparently undertaken, in order to counter threats that are far away from having acquired significant proportions. It is therefore, the case that Pakistan's introduction of battlefield nuclear weapons seeks to convey a destabilising effect and thereby, derive the effects that it has sought from its nuclear weapons even as the threat is incredible. These derivatives are:

- To portray the India–Pakistan situation as being on the abyss of nuclear war and thereby, invite intervention of the international community to prevent India from applying its conventional military power.
- To neutralise India's conventional response by creating ambiguity about its nuclear threshold through the introduction of battlefield nuclear weapons and thereby, continue to use terrorism as a foreign policy tool behind the shield provided by nuclear weapons.
- To increase the irrationality factor to reinforce deterrence.

In essence, the geopolitical emotion that guides Pakistan's nuclear policy is fear that is induced by a larger neighbour who they believe, poses an existential threat. Consequently, a belief system marinated in fear, has shaped its nuclear behaviour and there is nothing to suggest that it could change, except as a consequence of jettisoning the notion that India seeks its demise.

CHAPTER 5

INDIA'S NUCLEAR DOCTRINE

India should aim not just at being powerful; it should set new standards for what the powerful must do.

—**Non-alignment 2.0**

India's journey towards becoming a nuclear weapons power has been a diffident one. A nation that achieved its independence in 1947 through the embrace of ahimsa (non-violence), took a long time to eventually evolve into a nuclear weapon power in 1998. Although a demonstrative 'peaceful' nuclear explosion was carried out in 1974, but India recessed its capability for twenty-four years before it finally declared its transformation as a nuclear power. The original motivation for nuclear weapons came from China going nuclear in 1964. Between 1964 and 1984, Indian political leaders refrained from any move to develop nuclear weapons. In 1989, Prime Minister Rajiv Gandhi gave the go-ahead to start the process for building a bomb. This decision was taken because of the explicit and implicit nuclear threats experienced during the India–Pakistan crisis of 1984. There was also a greater realisation that India's security would require nuclear weapons since China was already a nuclear weapons power and Pakistan seemed to be in the process of becoming one. So, in essence, seeking security in an anarchic world was the prime reason for India becoming a nuclear power. It still agonised over the decision and the delay being a reflection of a decision that was forced upon it, due to developments in the security arena that warranted giving up its basic revulsion to nuclear weapons. India finally decided to carry out nuclear weapons tests in 1998, when the Bharatiya Janata Party (BJP), which had for some time been advocating the line that India must possess nuclear weapons, was elected to power.

Immediately after the nuclear tests, India stated unequivocally that its nuclear weapons were meant to perform only the core deterrence function and were not to be used for war-fighting. India's doctrine thus embraced NFU. India's weapons are meant only for preventing a nuclear attack on the country, rather than redress any military imbalance nor compensate for any conventional military inferiority or serve any operational or tactical battlefield requirements.

India persevered for twenty-four years after its nuclear explosion in 1974 to work for universal nuclear disarmament and a nuclear-free world. This was premised on a belief that a nuclear weapon-free world would enhance our security and that of the rest of the world. Unfortunately, this view received much lip service and nothing else. Instead, the division of the world into nuclear haves and have-nots was sought to be made permanent in the 1990s. To India it became clear that the possession of nuclear weapons was necessary if its attempts to promote a nuclear weapon-free world were to have some impact.

The National Security Advisory Board (NSAB), headed by India's foremost strategist K Subrahmanyam, prepared a draft nuclear doctrine which was released by Brajesh Mishra, the national security advisor on 17 August 1999[1] (Appendix A and B).

The draft doctrine describes the context in which India has sought to develop nuclear weapons. An environment of durable peace coupled with insurance against potential risks to peace and stability is considered as an essential prerequisite for the achievement of India's primary objective of economic, political, social, scientific and technological development within a peaceful and democratic framework. With nuclear weapons constituting the gravest threat to humanity and the virtual abandonment of nuclear disarmament by nuclear weapon powers, India seeks to exercise its right to self-defence through development and deployment of nuclear weapons with the fundamental purpose confined to deter the use and threat of use of nuclear weapons by any state or entity against India and its forces.

The doctrine states that India will not be the first to initiate a nuclear strike, but will respond with punitive retaliation should deterrence fail. A concept of credible minimum deterrence (CMD) coupled with 'retaliation only' is adopted and, therefore, survivability is considered critical. CMD is considered as a dynamic concept related to the strategic environment, technological imperatives and the needs of national security. Any nuclear attack on India and its forces shall result in punitive retaliation with nuclear weapons to inflict unacceptable damage to the aggressor. India will also not resort to the use or threat of use of nuclear weapons against states which do not possess nuclear weapons or are not aligned with nuclear weapon powers.

Sufficient, survivable and operationally deployed nuclear forces; robust command and control system; effective intelligence and early

---

[1] Draft Report of National Security Advisory Board on Indian Nuclear Doctrine, 17 August 1999. www.mea.gov.in/in-focus-article.htm?18916/Draft+Report+of+National+Security+Advisory+Board+onIndian+Nuclear+Doctrine

warning capabilities; comprehensive planning and training; the will to employ nuclear forces and weapons are identified as the imperatives for deterrence to prevail. The forces shall be based on a triad of air, land and sea vectors and capable of shifting in the shortest possible time from peacetime deployment to fully employable forces. Highly effective conventional military capabilities are considered necessary to raise the threshold of both conventional and nuclear war.

Nuclear forces and their command and control are to be organised for very high survivability against surprise attacks and for rapid response. Procedures for the continuity of nuclear command and control should ensure a continuing capability to effectively employ nuclear weapons. However, nuclear weapons are to be tightly controlled and released for use at the highest political level with the authority for use residing in the Prime Minister or his designated successor(s). An integrated operational plan or a series of sequential plans, predicated on strategic objectives and a targeting policy is expected to form part of the system. It also envisages the creation of space based and other assets for early warning and damage assessment.

Security and safety of nuclear weapons is expected to be maintained by extraordinary precautions that guard against theft, loss, sabotage, damage or unauthorised access or use. Tamper-proof measures are expected to be instituted to ensure unauthorised or inadvertent activation/use. A disaster control system is to be developed that is capable of handling accidents with nuclear weapons and materials.

Restraints on research and development capability will not be acceptable and efforts directed to keep up with technological advances will be made.

The doctrine concludes with a reiteration of its disarmament and arms control goals through the quest for a nuclear weapons free goal; efforts to persuade nuclear weapon states to join an international treaty banning first use and providing unconditional negative security assurances to non-nuclear weapons states; nuclear arms control measures and appropriate nuclear risk reduction and confidence-building measures shall be sought.

In the statement during the release of the doctrine on 17 August 1999, Brajesh Mishra highlighted some of the salient features of the doctrine. He said that nuclear weapons had become necessary due to the global security environment and the requirement of maintaining strategic autonomy in decision-making. India's nuclear weapons, he clarified, are not country specific and the policy of minimum but credible deterrence has been adopted as the basic building block of India's nuclear thinking and from which flows the NFU posture. He also emphasised the cardinal principle of civilian control.

The original draft nuclear doctrine of 1999 has never been explicitly and officially confirmed by the Cabinet Committee on Security (CCS) and there have been doubts raised in the public domain about whether or not the draft was actually approved. On 2 August 2000, the Minister of State for External Affairs, Ajit Panja, while replying to an un-starred question in the Parliament, clarified that 'the draft paper, Indian Nuclear Doctrine prepared by the NSAB, constitutes one of the inputs submitted to the National Security Council for finalising a doctrine and was released to encourage discussion and debate at wider public and political levels'.[2] But ipso facto it has been taken to be India's nuclear doctrine with some clarifications being officially announced in 2003 (Appendix C). The core criticism of the doctrine has focused on three main issues.

The first was regarding the adoption of NFU as one of the central pillars of the doctrine. This criticism was based on the argument that operationally it would be challenging to survive a first strike and strike back due to the possibility of damage to the nuclear arsenal and

[2] http://164.100.47.194/Loksabha/Questions/QResult15.aspx?qref=30401&lsno=13

its command and control systems. This criticism was also supported by the argument that India's military, which would be responsible for operationalising the arsenal, was never consulted. The supporting argument is valid, but it is irrelevant since India considered nuclear weapons for deterrence and not war-fighting and the military has no choice but to devise the optimum measures to ensure survivability which has been identified as a key operational capability that must shape the development of the nuclear arsenal. A bolt from the blue attack is the worst operational scenario that has the possibility of preventing nuclear retaliation. However, such a scenario is not only unlikely in the Indian strategic context but is in any case being tackled to a large extent by India building a SLBM capability, which is likely to consist of at least four-five nuclear-powered submarines within a decade or so. Though, when viewed purely in operational terms NFU is certainly disadvantageous, its political value is embedded in the idea that India views nuclear weapons as a necessity that arises from and is confined to its core deterrence function. NFU also negates the adversary's ability to politically leverage the nuclear card and coupled with civilian control, promotes stability.

Universally, the concept of NFU has never been popular with the military, fundamentally because, to forsake, willingly, the advantages of a first strike, is anathema to the military mind. The advantages that accrue from striking first is deeply ingrained in conventional military training and doctrine and is believed to maximise the chances of victory. It is another matter that history does not support this notion—the commencement of the First World War and Japan's attack on Pearl Harbour are obvious examples. The ultimate trigger for outbreak of wars is often a result of the political leadership being convinced by the military about the benefits of pre-emption based on an intelligence assessment of the adversary's intention. In the conventional arena, India's Cold Start doctrine is an archetypal product that is rooted in the military belief that 'striking first produces victory'. The idea of pre-emption with nuclear weapons cannot survive the political argument that between nuclear-armed rivals, a nuclear war cannot be won and that the first use decision by any of the actors, during a crisis or an ongoing conflict is virtually impossible to justify. NFU is the child of this logic apart from addressing the larger moral issue of mass genocide.

The first use of nuclear weapons in the pre-emptive mode based on perfect intelligence of the adversary's intentions is problematic in the real world. In his book *Choices* India's former National Security Advisor, Shivshankar Menon, notes that 'circumstances are conceivable in which India might find it useful to strike first, for instance, against a nuclear weapon state (NWS) that had declared that it would certainly use its weapons, and if India was certain that adversary's launch was imminent'. This has invited comments that seek to question India's adherence to NFU. In the scenario envisioned, one can presume that both countries have nuclear weapons on alert and there is an ongoing armed conflict. The scenario of perfect intelligence indicating an imminent launch is nearly impossible as the enemy's intentions will never be clear in such a context. Weapons could be deployed and be in the ready state, the signs of which may have been indicated by intelligence. But to establish beyond doubt the presumed intention of the adversary to launch, within a very narrow time frame, is virtually unviable, primarily because it would either lack corroboration or there would be diverse inputs whose veracity would be questionable. Intelligence in a situation of prevailing hostilities is increasingly fragile and more so because of complications caused by dual-use systems especially aircraft and missiles. Perfect intelligence does not exist in the real world especially in the midst of a situation where nuclear weapons are on alert in an ongoing state of armed hostilities. Uncertainty will reign supreme because it is simply the default condition which Clausewitz described as producing friction and fog of war. The stakes and consequences of making a misjudgement or miscalculation are astronomically high. Having rejected the idea of pre-emption unequivocally, India's nuclear doctrine has no exceptions to the NFU. The doctrinal emphasis on building a second-strike capability amply proves the point.

The second issue that faced criticism was the stipulation that 'nuclear retaliation to a first strike will be massive and designed to inflict unacceptable damage'. The NSAB draft had formulated this differently: 'any nuclear attack on India and its forces shall result in punitive retaliation with nuclear weapons to inflict damage unacceptable to the aggressor'. This change from 'punitive retaliation' to 'massive retaliation' was viewed as a prescription for reaction to be always confined to 'massive retaliation' and hence, lacked credibility especially if the aggressor has attacked with tactical nuclear weapons (TNW). Criticism of this issue has also intensified after Pakistan has announced its determination to deploy TNW. The scenario of Pakistan using a nuclear weapon on one of India's mechanised spearheads that has penetrated deep into Pakistan during an India–Pakistan war has underpinned the argument. The actual damage was assumed to be relatively limited and it might be used in Pakistani territory. Therefore, India's doctrine of retaliating to a level of causing 'unacceptable damage' through 'massive retaliation' was not only a case of disproportionate response that cannot be justified but also a response that could not be carried out and hence, lacked credibility. While the contention has been, that India should instead adopt a strategy of flexible response, the argument ignores some fundamental facts.

The idea that any use of a nuclear weapon could be tactical in effect is grossly misplaced. The use of a nuclear weapon would have a strategic and political impact with global reverberations. So, to view it as another military weapon for battlefield use is to ignore the actual implications of even what is supposed to be restricted nuclear use. The proponents of flexible response also believe that a strategic dialogue of some kind will prevail during the period of nuclear exchange, and therefore, one could, with carefully calibrated nuclear use, limit the damage to some acceptable level. This is a theoretical mirage that cannot possibly withstand the test of military realities. The friction and the fog of war will not permit any possibility of nuclear exchange being merely a strategic conversation based on a tit-for-tat retaliatory action and reaction. Even for conventional war, escalation is structurally intrinsic due to misperception, miscommunication, misjudgement and human frailties that arise from a situation characterised by fear and mistrust as the predominant emotions. More importantly, there is no guarantee that any proportionate reaction will be firstly viewed as proportionate and secondly, not invite a major response. It can be reasonably assumed that even the assessment of damage caused by a nuclear weapon would be extremely difficult to ascertain in the time frame required to shape a proportionate response. We must also assume that by this time the leadership would be deep inside the bowels of the earth in a protected nuclear facility and would be challenged to take decisions based on unconfirmed initial reports that would be contradictory and add to the uncertainty and confusion. India has, therefore, made it clear that it considers the strategy of flexible response as unworkable, impractical and one that would degrade our security. It would also, without doubt, drive up the numbers of our nuclear arsenal and become by itself a cause for greater insecurity.

The threat of 'massive retaliation' that is linked to 'unacceptable damage' also lends itself to different interpretations and provides leeway towards some degree of flexibility in our retaliation.[3] Both 'massive' and 'unacceptable damage' are better left unexplained to

---

[3]The author discussed with an official who preferred anonymity, involved in the CCS statement, the issue of change from 'punitive retaliation' in the NSAB draft to 'massive retaliation' in the CCS statement. No explicit reason was forthcoming except that perhaps it may have been felt that 'massive' would sound more menacing and therefore better deterrent effort.

provide whatever flexibility the actual context of the situation may demand. There is one factor on which there is no ambiguity whatsoever—India will undoubtedly retaliate, for to not do so could increase the possibility of suffering greater damage. After being attacked by nuclear weapons, this is not only politically correct but also an operational necessity. So, in the context of TNW, the doctrinal guidance is to plan for retaliation in a manner that would cause unacceptable damage which could also seek to neutralise the adversary's reaction capacity. It is only then that destruction and damages to India are minimised. In practice, use of TNW by Pakistan provides India with an option of striking second but without having experienced a substantive first strike and with its nuclear arsenal intact: this issue has also been flagged in Shivshankar Menon's book, *Choices*.[4]

The third issue was regarding the deterrence of chemical and biological weapons threats. The Draft Doctrine did not include the deterrence of chemical and biological weapons by nuclear weapons. But the CCS official statement in 2003 made it explicit that chemical and biological weapons were also sought to be deterred. It stated: 'However, in the event of a major attack against India, or Indian forces anywhere, by biological or chemical weapons, India will retain the option of retaliating with nuclear weapons.' So, chemical and biological threats were not only included but by stating that '*India will retain the option* of retaliating with nuclear weapons', it embraced a provisional clause, that kept at bay the 'commitment trap', which is a phenomenon that would commit us to a reaction even if it is a minor episode. So, for nuclear weapons, it is 'any use', for chemical and biological weapons it is 'retention of the option'. This is so because although nuclear, chemical and biological weapons are often clubbed in the same basket as weapons of mass destruction, their

[4] Menon, Shivshankar. op cit, p 174.

physical effects could have large degrees of variance. Chemical and biological weapons could also have restricted effects and therefore, a commitment to use nuclear weapons to deter all use of chemical and biological weapons runs the risk of being tested for reaction that cannot be strategically and politically justified. The phraseology 'retain the option' provides an escape from the commitment trap.

India's operational capability in the nuclear domain has been a subject of much conjecture. The main capabilities that have been projected are its missile reach which is likely to be the Agni-V with a range of 5,000 km and its development of nuclear-powered submarines that would in due course house a missile with a 5,000 km plus range. In April/May 2013, India successfully tested the K-4 missile with a range of 3500 km from an underwater platform. The Agni series I, II, III, IV and V are all solid fuel, road/rail mobile, indigenously designed and produced. There is, however, a requirement to build considerable redundancy in the command and control structure.[5] This would signal a determination that should be tangible enough for nuclear adversaries to take note of.

As deterrence is an inherently unreliable method with the possibility of Pakistan's nuclear arsenal falling into the hands of jihadi elements, India is, therefore, making robust efforts to deploy missile defence systems. A process that is underway and according to the head of India's Defence Research and Development Organisation (DRDO), is in a very advanced stage of development.[6] However, it is this very development of missile defences by India that runs the risk of being the catalyst in pushing its adversaries to expand their nuclear inventories and provide the momentum for an arms race. However, India's nuclear doctrine being based on a credible minimum deterrence philosophy could minimise the possibility if

---

[5] ROD with Bruce Blair, 'Even one nuclear weapon will require a fairly complex command and control system', Washington DC, 21 February 2001.

[6] https://en.wikipedia.org/wiki/Indian_Ballistic_Missile_Defence_Programme

the rationale for missile defences development is restricted, at least initially, to protecting only India's command and control elements in New Delhi, primarily against rogue elements who may take control of Pakistan's nuclear weapons. Subsequent to the deployment of such capability and drawing on that experience the role of India's BMD could be reviewed vis-à-vis threats envisaged.

The development of MIRVs by India would seem inevitable, considering similar development in the neighbourhood which is also accompanied by ongoing efforts to develop and deploy BMD systems. The essential principle to guide MIRV development and deployment should be that it should not inflate nuclear weapon quantities. Therefore for India, MIRVs based on 5,000 km missile SSBN fleet should be the natural choice as it can simultaneously shrink its air and land vector systems.

Considering the level India's satellite and missile launch capability has already attained, hypersonic vehicles would be a natural choice of development for long-term benefits. It would also depend on the availability of financial resources as also the feasibility of miniaturisation of warheads through laboratory tests. As yet, there are no indications in the open domain of any such development project.

India has fashioned its doctrine on the basis of a categorical acceptance that nuclear adversaries could resort to nuclear weapons 'in extremis' and has adopted the concept that it is the survivability of the arsenal that matters much more which is of greater significance than the size of the nuclear arsenal per se. A compact, mobile, opaque arsenal based on road, rail, underwater and airborne platforms is appropriate for India and, with the complete induction of nuclear submarines, India's nuclear arsenal's survivability would reach a high level of self-assurance. The defiance of the US and the international community by North Korea in exhibiting its missile and nuclear capability in 2017 provides sufficient evidence that even a country with a nascent nuclear capability based on mobile launchers could deter the most militarily powerful country from using its conventional and nuclear capability. It merely goes to offer proof that successful nuclear deterrence does not lie in numbers but in creating that element of doubt that maybe one or a few nuclear weapons could survive and visit destruction on the attacker.

In the last decade or so, while India has made tremendous advances in operationalising its nuclear wherewithal, it has intentionally adopted a policy of not giving publicity to developments in its operational capacity, which has given rise to much uninformed criticism of India's nuclear weapons development. On the other hand, it is an arena that has made progress quietly, steadily, significantly and purposefully. Though based on hearsay, it probably has a very ingenious system to counter a surprise decapitation strike. The nuclear doctrine in its original form has guided the growth of India's arsenal. Notably, the growth has been at a steady pace and without entanglement in an arms race. Admittedly, it has some distance to travel in terms of being based on a triad. The geopolitical emotion that has guided India's nuclear policy is that of hope which finds its ultimate expression in a nuclear doctrine that rests on the belief that the only role of nuclear weapons is to deter their kind.

CHAPTER 6

KARGIL: DETERRENCE PERSPECTIVES

Pakistan's military establishment had entertained ideas of deterring Indian nuclear and conventional capabilities with its nuclear weapons and carrying out a brash, bold strike to liberate Kashmir which would go unchallenged if the Indian leadership was weak and indecisive.

—**The Kargil Committee Report**

Kargil is separated from the Kashmir Valley by the great Himalayan range and is accessible through National Highway 1-A (NH-1A) via the Zoji La pass, which is snowbound from December to April. The Line of Control (LC) comes uncomfortably close to NH-1A as it emerges out of Zoji La and dips southwards close to Drass at a height of 5,353 m, from where it runs almost parallel to the NH-1A, till Kargil town. In mid-1999, the LC in proximity to Kargil played host to a limited India–Pakistan conflict.

Pakistan had consistently targeted the road in the Kargil–Drass area by artillery fire, to hinder the movement of convoys proceeding from Zoji La towards Leh. In the 1965 war, major battles were fought with an aim to remove Pakistan's direct observation over this area. The Pakistani posts at Pt 13,620 (ft) and Black Rock (15,000 ft) were captured, but handed back under the Tashkent Agreement and during the 1971 War, these posts were again recaptured along with the contiguous areas of Tortuk and Chalunka in the adjacent Shyok Valley.

The alignment of the LC in Kargil has never been in dispute and east of Kargil and Batalik, the LC was never physically held in strength though some special patrols operated in these areas even in winter. No major offensive thrust could be launched, because of unfavourable weather conditions and a long and arduous logistic lifeline via the rugged terrain of Gilgit and Skardu. Interdiction of convoys by artillery fire or temporary blockages by infiltrating columns could cause delays but do no more. To capture areas around Drass or Kargil or anywhere near the NH-1A and sustain it logistically was assessed to be beyond Pakistan's military capability.

Planning

Exasperated by the Indian artillery fire and sniping into the Neelam Valley in Kashmir, as well as the Indian actions on the Siachen Glacier, Pakistani military tacticians looked for a site where India could be given a taste of their own medicine. Along most of the LC, the Indians held the dominating heights so only a few places looked suitable. The most obvious site was the Kargil-Drass sector. The knowledge that every winter the military posts in the area were vacant, and the realisation that the crucial NH-1 would be easy to interdict, made it more suitable.

The Pakistan Army's plan was to occupy the dominating heights in the Kargil-Drass area while Zoji La pass was closed and before the Indian Army reoccupied the positions, using regulars disguised as Kashmiri militants. They assumed that the Indians needed time to assemble their forces, acclimatise their troops, build up their logistics, which would be difficult before the Zoji La pass opened and with the advantage of the commanding heights, better acclimatisation and logistics, the situation would be distinctly in their favour. India would also have to suffer unacceptable casualties in attacking the heights which they calculated, would provide sufficient time for an internationally arranged ceasefire.[1] The initial plan was executed successfully and the un-held heights were occupied by Pakistan army regulars in the guise of militants by April 1999.

It seems that there were some major purchases of winter military equipment by Pakistan and that certain units with the responsibility for the defence of the LC had been reinforced through 1998-99. According to intelligence sources, the number of mujahideen being trained in north-east Pakistan increased markedly towards the end of 1998. Around forty new camps sprang up in the Neelam Valley, around Muzaffarabad and in the barren hills of Baltistan. When a reporter visited Skardu in April 1999, bearded fighters in combat kit, many speaking Pashto, the language of the North-West Frontier Province and Afghanistan, were very much in evidence in the bazaar.[2]

What is beyond doubt now, however, is that at some point in the winter of 1997-98, Pakistan's armed forces and the Inter-Services Intelligence (ISI) deliberately chose not to accept the status quo in Kashmir and opted, instead, for conflict. It is possible that the Nawaz Sharif government might have been kept in the loop then with the caveat that these were war games and contingency plans, but as the pressures of bankruptcy intensified after Pakistan's nuclear explosions, and Sharif turned to the International Monetary Fund (IMF) for a bailout, the army headquarters must have concluded

---

[1] The Kargil Review Committee Report, '*From Surprise to Reckoning*', Sage Publications, New Delhi, 2000, p 229.

[2] Burke, Jason. 'Kargil War: Pakistan—In the Land of the Enemy', *India Today*, 12 July 1999, p 20.

that the civil government could no longer be relied upon. From then on, Kargil seemed purely a military initiative,[3] though General Musharraf has continued to insist that everyone was on board the Kargil operation. The chairman of India's officially sponsored Kargil Committee Report, K Subrahmanyam, has been quoted as affirming that Sharif was well aware of the planned incursions and had visited forward positions on the Pakistani side days before the Lahore Summit.[4] Sharif was deposed by President Musharraf in October 1999 and convicted to a life term. During his deposition at the trial, referring to the fact that his differences with Musharraf dated back to Kargil which was one of the worst crises in the country's history, the former Prime Minister said, 'Musharraf was playing hide and seek on this issue not only with me but also with the armed forces. My objection to this conduct became a source of resentment for him…I will not for reasons of national security elaborate any further on this issue, although it is necessary for the people of Pakistan to be informed of about the truth about Kargil.'[5] He later told reporters at Attock: 'This ill-planned and ill-conceived operation was kept so secret that besides the Prime Minister, the chiefs of the navy and air force were also kept in the dark.'[6] The Musharraf government spokesman responded to Nawaz Sharif's statement by stating that Sharif had been briefed at least four times on the military aspects of the Kargil operation.[7]

India's Reaction

The Indian Army detected the intrusions between 3 and 12 May 1999 and military action to evict the intruders commenced around

[3] Jha, Prem Shankar. 'The Mind of the Pakistan Army', *The Times of India*, 16 June 1999.

[4] Subrahmanyam, K. quoted at the release ceremony of the Kargil Committee Report, *The Times of India*, 25 February 2000.

[5] Baruah, Amit. 'Whose Kargil Version to Believe', *The Hindu*, 14 June 2000, p 14.

[6] Ibid.

[7] *Daily Jang*, Lahore, 15 March 2000.

26 May. The mission was code named Operation Vijay. The Indian Army reacted much faster than what Pakistan had anticipated and the Indian government authorised force application without violating the LC on 25 May 1999. While the tactical battle was being fought in Kargil, steps were taken at the strategic level to raise the ante for Pakistan and to influence the course of events through reinforcement of the Western Fleet of the Indian Navy from the Eastern Fleet and the forward movements of strategic formations of the Indian Army towards the border in the western and southern theatre. Pakistan, meanwhile, tried to lobby with the international community for a ceasefire, which would leave it with some parts of Indian territory, and, therefore, justify its misadventure. Initially, there was support for a ceasefire but once the Indian military success unfolded, the international community called on Pakistan to withdraw from and respect the sanctity of the LC. India succeeded by July 1999 in forcing Pakistan to withdraw from Indian territory, albeit at the costs of substantial military casualties and intervention by the US.[8]

Pakistan's Motivations

The Kargil Review Committee Report cites Pakistani writings to reveal the likely motivations for undertaking the Kargil intrusions with the prime politico-strategic motive cited as 'to internationalise Kashmir as a nuclear flashpoint requiring urgent third-party intervention'.[9] The report further postulates that Pakistan had likely undertaken the Kargil intrusions based on the assumption that their nuclear capability would forestall any major Indian conventional response; it also gambled on the hope that the international community would

[8] Dixit, JN. 'Pakistan Machinations in Kargil', *Hindustan Times*, 7 July 1998. Dixit states the major reason for the US mediation was the fact that the conflict could escalate into the nuclear realm.

[9] *From Surprise to Reckoning: The Kargil Review Committee Report*, Sage Publications, New Delhi, 2000, p 89.

intervene early thus enabling Pakistan to bargain from a position of strength.

Pakistan's actions also undermined the sanctity of the LC that has served as a de facto border between the two nations in Kashmir for more than a quarter of a century. Pakistan sent Foreign Minister Sartaj Aziz to India in early June 1999 for talks and his principal objective in New Delhi seemed to be to develop the case that talks with India were futile and demand the intervention of the international community so as to compel India to negotiate under pressure. Projecting a failure on the diplomatic front and escalating the conflict in Kashmir were the two elements of Pakistan's strategy to challenge the very meaning of the LC and seek international intervention.

Stability–Instability Paradox

India weaponised its nuclear capability between 1992 and 1994.[10] The development had been underway since the late 1980s and former chief of air staff told the Kargil Review Committee that he was inducted to work for the nuclear delivery programme sometime in 1986. It is therefore possible that Pakistani threats between 1984 and 1990 were founded on the notion of nuclear asymmetry, a topic that was being widely debated in India during that period.[11] Ever since the demonstration and declaration of nuclear capability by India and Pakistan in May 1998, the overt situation had changed. Though both sides possessed only rudimentary arsenals, the mere possibility of suffering destruction on a catastrophic scale was considered by Pakistan as sufficient to deter an Indian conventional

---

[10] Ibid, p 205.

[11] Sundarji, K. *Blind Men of Hindoostan*, UBS Publishers Distributors, New Delhi, 1993. In a mix of fact and fiction, Sundarji demonstrated the folly of nuclear asymmetry in the India–Pakistan context. See also, 'Effects of Nuclear Asymmetry on Conventional Deterrence', *Combat Papers*, No. 1, The College of Combat, Mhow, 21 May 1981. In contrast, the present debate is focused on the application of conventional power in a situation of nuclear parity.

response to Pakistan's proxy war in Kashmir, and more importantly, to call international attention to the image of South Asia being a nuclear flashpoint. However, the thinking in India on conflicts in the nuclear era has been focused on the issue of limited war, which does not quite fit into the Pakistani paradigm of the nuclear umbrella facilitating a proxy war without inviting an Indian response.

The origins of the stability–instability paradox lie in the tension that nuclear weapons have generated between political objectives and military force application. Victory through use of nuclear weapons between nuclear-armed adversaries is no longer possible and conventional war possibilities are now endangered due to the fear of escalation into nuclear realm. In the India–Pakistan context, a nascent nuclear arsenal coupled with an ongoing proxy war has presented both countries with a strategic dilemma that reflects the end products produced of the stability–instability paradox.

Pakistan's strategy has been to utilise the political value of nuclear weapons by using the nuclear situation to paint South Asia as a nuclear flashpoint. At the same time, it feels emboldened that its proxy war will not invite an Indian conventional retaliation due to the presence of nuclear weapons. For the past several decades, Pakistan has been fighting a proxy war with India using militants trained, equipped and operating under their orders with terrorism being their chief instrument and has included a fairly large percentage of Pakistanis. Pakistan's strategy to wrest Kashmir by force is no longer a possibility especially due to the emergence of the nuclear factor between the two. But they believe, the nuclear factor can facilitate political victory if the nuclear card is used to secure international intervention. How international intervention will facilitate political victory for Pakistan is itself a highly debatable issue and is one more illusion that Pakistan has nursed over the years. The Kargil adventure was also a reflection of this illusion especially when the nuclear capability of both nations was still at a nascent stage.

Pakistan's Nuclear Capability

Historical evidence of Pakistan's growing involvement in Jammu and Kashmir suggests that its involvement growth curve trajectory paralleled its nuclear capability[12] which at the time of Kargil consisted of a very limited but unknown number of nuclear warheads. Its delivery capability restricted primarily to a few aircraft (F-16s or Mirages) suitably modified. Though Pakistan had earlier carried out several tests of short-range (Hatf and Shaheen) and medium-range (Ghauri) missiles, none of them were operational at the time of the Kargil War. In summary, Pakistan at the time of the Kargil Conflict possessed an opaque and highly limited quantity of un-deployed nuclear weapons and restricted delivery capability. The arsenal by former President Musharraf's own admission was not operational.[13] It was with a non-existent capability around which Pakistan wove its paradigm of a nuclear umbrella, to prevent escalation as well as leverage the nuclear flashpoint card internationally.

Pakistan's Beliefs and Nuclear Threats

The roots of the Pakistani belief that nuclear deterrence will prevent India from any major conventional reaction lay in their misperception that they had successfully deterred India through nuclear threats in 1984, 1987 (Brass Tacks) and the 1990 crises.[14]

In 1984, the Pakistani government threatened to retaliate with 'all the means at its disposal', though the use of nuclear weapons was neither specifically mentioned nor excluded.[15] This threat, if

[12] The correspondence of the acquisition of nuclear capability and the prosecution of war is illustrated in Jasjit Singh, 'Kargil 1999: Pakistan's Fourth War of Kashmir', *Strategic Analysis*, Vol-13, No. 5, IDSA, August 1999, p 691.

[13] Musharraf, Pervez. *In the Line of Fire*, Simon & Schuster, UK, 2006, pp 97-98.

[14] Shahi, Agha; Khan, Zulfiqar Ali and Sattar, Abdul. 'Securing Nuclear Peace', *The International News*, 5 October 1999.

[15] Ibid.

intended as a nuclear one, was an outright bluff as Pakistan probably achieved nuclear weapons capability only in 1987.[16]

In 1987, during Operation Brass Tacks, when Pakistani and Indian forces faced each other, SK Singh, the Indian ambassador in Islamabad, was summoned to the Pakistan Foreign Office and Zain Noorani, the Pakistani minister of state for foreign affairs, asked him to convey a message from General Muhammad Zia-ul-Haq, 'If India took any action not conducive to its sovereignty and territorial integrity, then Pakistan was capable of inflicting unacceptable damage on it. Pakistan's action would not be limited to northern India alone but also to facilities outside the north.' When asked whether this implied an attack on Mumbai, the Pakistani minister replied that it 'might be so'.[17] A nuclear threat was also publicly conveyed by Dr Abdul Qadeer Khan, Pakistan's chief nuclear scientist, in an interview published after the end of the military exercise in the British newspaper, *The Observer*.

In 1990, when India–Pakistan tensions were high, Pakistan's foreign minister, Yaqub Khan visited India from 21-23 January and met the then external affairs minister, IK Gujral and Prime Minister VP Singh. Gujral in his meeting with the Kargil Review Committee recalled that Yaqub Khan was very melodramatic and was reading from a prepared note and used terms (in Urdu) like 'fire would be spitting from the skies and there would be a holocaust'.[18]

Pakistan's beliefs of having deterred India had been based on the notion that India was preparing or planning an attack. This was, in fact, not the case and such an option was not available or seriously considered due to the commitment of the Indian Army in Sri Lanka,

---

[16] The Kargil Review Committee Report, op cit, p 191. The report quotes General Zia and an Indian Intelligence report of March 1988 that estimates that Pakistan was in possession of at least three nuclear devices of 15-20 kt yields.

[17] Ibid, p 191.

[18] Ibid, note 17, p 79.

countering terrorism in Punjab and insurgency in the North East and in Jammu and Kashmir.[19] In psychological terms, Pakistani fear psychosis conjured imaginary fears, which they thought they had successfully countered through nuclear threats and found security in the belief—a belief which has endured and from the Pakistani viewpoint been proven in the 2001-02 crisis and the Mumbai terrorist attack in 2008 and continues to be the source of their miscalculation in future crises.

India–Pakistan Situation: Competing Bluffs

The India–Pakistan nuclear threats are fundamentally unbelievable because of the possibility of mutual destruction. So, when Pakistan promotes the notion that its nuclear capability will be used early in a conventional conflict, what it logically means is that, at that point, Pakistan has decided to commit suicide. It is an obvious military bluff meant for political effect.

India, on the other hand, threatens nuclear retaliation only to the use of nuclear weapons, which it describes as 'massive and designed to inflict unacceptable damage' and postulates the feasibility of limited conventional war. Limited war objectives are inherently discordant with large penalties and to risk it all for modest objectives appears senseless, especially if the penalty is not credible; risk-taking by one side will likely prompt risk-taking by the other. Neither adversary, as Robert Jervis has written, 'can confidently move from one area of significant concern to the other without great risk of incurring very high costs—if not immediately, but as a result of a chain of actions that cannot be entirely foreseen or controlled'.[20]

The critique of massive retaliation by Henry Kissinger and other Cold War deterrence strategists still rings true, 'Given the

---

[19]The Kargil Committee Report, ibid, p 76 and note 44, p 80.

[20]Jervis, Robert. *The Meaning of the Nuclear Revolution*, Cornell University Press, New York, 1989, p 32.

power of modern weapons, a nation that relies on all-out war as its chief deterrent imposes a fearful psychological handicap on itself. The most agonising decision a statesman can face is whether or not to unleash an all-out war; short of a direct attack threatening the national existence, all pressures will make for hesitation, short of a direct attack threatening the national existence.'[21]

Pakistan's reliance on threats based on escalation to the nuclear realm to deter India's application of conventional force is, therefore, considered by India, to be a bluff that should be called whenever required. Importantly, the point to note is that such dilemmas do not apply to a NFU power like India.

Lessons for Deterrence

Nuclear Factor

In any future war situation, Pakistan will flash its nuclear card as early as possible in order to invite international intervention and neutralise India's conventional edge. India's reaction would depend on the stakes involved. The nuclear card, though fundamentally incredible, can carry political utility and have strategic effect. Here the engagement takes place between the minds of the opposing leaderships. Therefore, even if nuclear weapons are not actually used, the outcome of the political crisis or even the political effects of limited military engagements are influenced by the nuclear factor. Bruce Riedel, who was a special assistant to American President Bill Clinton, has written that he was present when the President informed Sharif that the Pakistani Army had mobilised its nuclear-tipped missiles.[22] India's foreign minister at the time of the Kargil crisis, Jaswant Singh, confirms that, 'India had some information regarding some deflective activity in Pakistan's Tilla ranges near Jhelum, indicating

[21] Kissinger, Henry. *Nuclear Weapons and Foreign Policy*, Harper, New York, 1957, p 133.

[22] Reidel, Bruce. 'American Diplomacy and the 1999 Kargil Summit at Blair House', Policy Paper, Centre for Advanced Study of India, University of Pennsylvania, 2002.

that it could be operationalising its nuclear missiles. This was treated by India as a mere gambit. A nuclear angle to this conflict simply did not exist.'[23] Singh, however, elucidates that any decision other than the one to keep the conflict geographically confined in spatial terms and not cross the LC though 'tactically disadvantageous, would have been a strategic error of incalculable dimensions, principally because of the nuclear status of both India and Pakistan, which for the world was the primary worry. We had assessed this carefully and were clear that there was to be no internationalising of the issue; we wanted no repeats of Tashkent; we undertook, therefore, diplomatic manoeuvres to ensure that the matter did not go before the Security Council.'[24] General VP Malik, Indian Army chief during the Kargil War, also confirms that the 'nuclear weapons factor played on the minds of the political decision makers' and confirms that the nuclear issue was not discussed in the CCS, the highest political decision-making mechanism, though it was discussed between General Malik and the National Security Advisor Brajesh Mishra a couple of times.[25]

There is, therefore, sufficient grounds to believe that even though nuclear weapons were never physically used in the Kargil crisis, the political shadows cast by the nuclear factor influenced Pakistan's planning, India's reaction and its eventual conclusion through the diplomatic intervention of the US, which also points in the direction of the need for international support for waging wars.

International Support

India's decision not to violate the LC was a decision largely aimed to garner international support in the crisis. This was in recognition of the belief that international support was a strategic requirement in waging modern wars as there are limits to national sovereignty in

---

[23] Singh, Jaswant. *India at Risk*, Rupa Publications, New Delhi, 2013, p 203.

[24] Ibid, p 203.

[25] Malik, Gen VP. *India's Military Conflicts and Diplomacy*, Harper Collins, India, 2013, p 127.

undertaking military operations, even in its own defence.[26] They not only add a new perspective to the nature of future wars and forces required, but that international support and cooperation, form essential ingredients to waging conflicts successfully.

Escalation Control

The Kargil War is considered illustrative of a conflict in which escalation was successfully controlled. The peculiarities of the Kargil War are, however, noteworthy. Pakistan sought to disguise aggression through a façade of mujahideen having carried out the operation.[27] India reacted by announcing its policy of not crossing the LC, simultaneously signalled its resolve to escalate by the forward movement of its strategic formations, and movement of naval ships. The Kargil Review Committee Report concludes that these moves deterred Pakistan from escalating the conflict.[28] But this conclusion is founded on the assumption that Pakistan intended to escalate the conflict and was prevented from doing so. Whereas from all accounts of Pakistani objectives and the intercept of the telephone conversation between Musharraf and Muhammad Arif, the chief of general staff, suggest that Pakistan had no intention to escalate and had asked their air and ground people to stay back[29] and the Pakistan Air Force did not react to the use of air power within Indian territory, though the Indian Air Force chief had initially expressed the view that the use of air power would escalate the conflict.[30] The Pakistani Navy also kept well away from Indian naval deployments. What the Indian moves probably achieved, as the Kargil Review Committee Report states,

---

[26] Raghavan, VR. 'Strategic Pointers from Kargil', *The Hindu*, 27 June 2002. See also Singh, Jaswant. op cit p 203.

[27] The Kargil Review Committee Report, op cit, p 97.

[28] Ibid, p 105.

[29] Transcript of telephone conversation, *The Hindu*, 11 June 1999–'Now the air people and the ground people will stay back and the situation will be OK.' See also, full transcript, Singh, Jaswant. op cit, pp 190-97.

[30] Kargil Report, op cit, p 232.

is that it sent a clear message that all intruders will be evicted at any cost. For India, the stakes were high enough to escalate, if necessary. Deter, it probably, did not, because the Pakistani game plan was to use the clandestine nature of the intrusion to prevent escalation.

Nuclear Linkage

During the Kargil Conflict, nuclear signalling by Pakistan was restrained. This appears to have been due to Indian military action being politically limited to its own side of the LC. The Kargil Review Committee Report reveals that Pakistan conveyed some veiled nuclear signals to India during the conflict. On 30 May 1999, soon after India employed air power in the Kargil Conflict, the Pakistan foreign secretary, Shamshad Ahmad told *The News International* and *Daily Jang* newspapers: 'We will not hesitate to use any weapon in our arsenal to defend our territorial integrity,' given the overt nuclearisation of Pakistan in May 1998, this threat had obvious connotations.[31] The statement was denied the same night by the Pakistan foreign spokesman by saying that the foreign secretary had been completely misquoted and his comments reported out of context. The nuclear threat, however, made international headlines and may have served, albeit marginally, the Pakistani purpose of linking the war to the nuclear issue. The important nuclear lesson of Kargil was that such linkages would be easier in future conflicts.

Political Caution

While India's political decision not to cross the LC was conveyed unequivocally, it's rationale for not crossing the LC has been interpreted differently. The official Indian view is that it was aimed to discredit Pakistan, bolster India's image as a responsible nuclear power and garner international support. Pakistan, on the other hand, believed that India's decision reflected the impact of nuclear deterrence and

---

[31] Baruah, Amit. 'We will not hesitate to use any weapon in our arsenal to defend our territorial integrity.' *The Hindu*, 31 May 1999.

bolstered their image of India as a 'soft state'. One could postulate that things would have been different if nuclear weapons were non-existent, a view that is buttressed by the historical fact that India reacted differently in 1965. But the more important point to note is the fact that the boundaries of the conflict were circumscribed politically, communicated to the adversary and respected by both sides—a reflection of a cooperative endeavour. It would, however, be imprudent to expect such cooperation during future conflicts.

Crystallisation of Belief Systems

Pakistan believed that nuclear weapons had constrained Indian response options, and therefore, they can successfully use nuclear deterrence to support aggression and a proxy war. They felt that the threshold of conflicts could be raised and had successfully taken greater risks. Indians believed that a limited war against Pakistan could be fought and won despite the presence of nuclear weapons. This clash of belief systems represented the potential for a future war. Both sides imbued deterrence with different potentialities. These beliefs sustained and laid the foundations of a potentially dangerous situation and the eruption of the India–Pakistan crisis of 2002 tested these belief systems.

CHAPTER 7

DETERRENCE REVISITED: INDIA–PAKISTAN CRISIS 2001-02

Once on the tiger's back, we cannot be sure of picking the place to dismount.

—**George Ball**

A suicide attack on the Jammu and Kashmir state assembly by Pakistan-based Jaish-e-Mohammed (JeM) left scores dead on 1 October 2001. India reacted by stepping up artillery attacks on Pakistani military posts on the LC in order to increase military pressure and complemented it by a concurrent diplomatic campaign to highlight Pakistan's sponsorship of terrorist activities.

On 13 December 2001, when terrorists attacked the Indian Parliament in New Delhi, Prime Minister Atal Bihari Vajpayee addressed the nation: 'Parliament is the highest representative body of democracy and they chose Parliament deliberately as their target... we accept the challenge and we will blunt every attack.'[1] This attack, for which a Pakistan based militant group was suspected, outraged the

---

[1] *Hindustan Times*, 13 December 2001.

Indian government and its populace, elicited global condemnation and persuaded the US to ban the Lashkar-e-Taiba (LeT) and JeM. The magnitude of the incident forced the Indian government to alter the terms of engagement with Pakistan and threaten Islamabad with the prospects of war with India mobilising its armed forces and Pakistan followed suit.

On the political front, India downgraded diplomatic relations by recalling its ambassador, cutting its consular strength and placing similar restrictions on Pakistan. India declared that all land and rail links with Pakistan would be suspended from 1 January 2002 and flights of Pakistani aircraft through Indian airspace were banned. By early January, borders were mined and Indian strike formations deployed. Menacing war clouds emerged on the horizon.

Prime Minister Vajpayee wrote to the heads of the governments of the US, Britain, Germany and France that Pakistan must understand that there was a limit to the patience of the people of India and India might be forced to take the ultimate step.[2] The letter implied the distinct possibility of war. The diplomatic and military build-up was meant as a signal to Pakistan that India's threshold of tolerance had been breached. A strategy of coercive diplomacy[3] seemed to be underway[4] that was also aimed at the US to apply pressure on Pakistan, to commit, to a visible and definitive end to cross-border terrorism.

---

[2] *India Today*, 3 June 2002, p 28.

[3] Coercive Diplomacy is the use of threat of force in aid of persuasion. See George, Prof Alexander. *Forceful Persuasion: Coercive Diplomacy as an Alternative to War*, US Institute of Peace Press, 1991, Washington, p 44. He postulates that two rules govern the game—the opponent must be allowed to save face by offering him a line of retreat; and the practitioner must have a fall-back position for himself or an exit strategy before rushing into a confrontation. It follows that the lines of communication must be kept in good repair.

[4] Singh, Jaswant. *India at Risk*, Rupa Publications, New Delhi, 2013, p 249.

Initial Crisis Eases

Succumbing to India's coercive diplomacy and exertions of US State Department, President Musharraf in an address to the Pakistani people on 12 January 2002, spelled out a sweeping domestic reform agenda. He declared the government's resolve to detoxify Pakistani society of the poisonous influence of militant, sectarian and radical Islam and build a more humane, secular and liberal alternative. To achieve the objective, the Pakistan government banned militant and extremist Islamic groups involved in sectarian campaigns and their offices were sealed, leaders arrested and funds frozen.

Musharraf's stated intentions temporarily defused the crisis but India ruled out demobilisation of its armed forces until such time as the Indian government was convinced that Pakistan had indeed reduced support to cross-border terrorism. In the interim, India demanded that Pakistan turn over twenty individuals accused of terrorist crimes committed in the country and decided to continue with a wait-and-watch policy while forces on both sides remained in an eyeball-to-eyeball situation. At that time, some Indian analysts believed that having got Pakistan to make promises, the deployment had achieved its aim and also that in geopolitical terms, war was not an option, as it was inconceivable to them that an Indian government would contemplate war without the minimum international empathy.[5] Perhaps, it may also have been the case that India's threat worked because by getting the Americans into action, they convinced Musharraf to issue the statement.

Second Crisis

Just when it seemed that the India–Pakistan military stand-off had begun to show signs of de-escalation, tensions flared up again in response to terror attacks at Kaluchak in the Jammu region by

[5] Raghavan, VR. 'Military Persuasion or Provocation', *The Hindu*, 12 February 2002.

a Pakistan-based insurgent group on 14 May 2002. Musharraf claimed that his government lent only moral and political support to disaffected Kashmiri militants fighting Indian security forces in Kashmir. Exasperated with Pakistan's mendacity, the Indian government threatened war.

In turn, Pakistan expressed alarm at the threat of war and warned India of their ferocious response. Musharraf threatened that if attacked, Pakistan would take the war into the Indian territory. More ominously, Pakistani government representatives made it plain that if pushed into a corner by its larger and more powerful neighbour, Pakistan would not hesitate to use its nuclear arsenal in self-defence.[6] To reiterate this point, Pakistan conducted a succession of ballistic missile tests in the last week of May 2002. Although the tests were aimed primarily for the domestic audience, they once again highlighted the dangers of larger conventional war in South Asia and the nuances of nuclear signalling.

Nuclear Signalling[7]

The ten-month border confrontation (December 2001–October 2002) was the longest period of military mobilisation by both countries since their independence in 1947. A variety of nuclear signals took place, flight tests of ballistic missiles, press briefings, speeches to the public and the armed forces. With the disruption of diplomatic channels, public statements were the main means of communication and emanated from multiple levels in both the countries—political, military and bureaucratic.

---

[6] Shourie, Dharam. 'Defiant Pakistan Threatens to Use Nukes,' Rediff on the Net, 30 May 2002.

[7] This is a compilation based on Chaudhury, Rahul Roy. 'Nuclear Doctrine, Declaratory Policy and Escalation Control'. https://www.stimson.org/content/nuclear-doctrine-declaratory-policy-and-escalation-control

To distil aims and understand the messages embedded in nuclear signalling, two phases of the confrontation are being examined. The first phase can be defined as the period between the terrorist attack on Parliament on 13 December 2001 and prior to Kaluchak on 14 May 2002; the second phase covers the Kaluchak period until the easing off of the crisis in mid-June 2002.

Phase I: 13 December 2001-13 May 2002

These five months cover the period when India came close to using force against Pakistan. India appeared keen to send two major signals to multiple constituencies—its domestic public, the Pakistani government and the US. First, that its much-publicised threat to use conventional force against Pakistan was real and credible, with limits to its restraint and patience fast depleting, unless, of course, Pakistan complied with its demands to end cross-border terrorism. The second was that it would strenuously avoid any nuclear signalling to Islamabad, as well as deliberately ignore any nuclear signalling from Islamabad.

In fact, on 1 January 2002, within two weeks of India's mobilisation, New Delhi and Islamabad exchanged navigational coordinates of their nuclear installations and facilities, as they had done since 1998, in accordance with the bilateral agreement on the Prohibition of Attack against Nuclear Installations and Facilities. Apparently, neither New Delhi nor Islamabad felt it prudent to discontinue the existing practice on this confidence-building measure (CBM), notwithstanding tense bilateral relations. However, this may have sent mixed signals to each other of reassurance that nuclear-related CBMs were insulated from the upheavals of politics, although this may not actually have been the case.

Not only was Musharraf's speech on the occasion of Pakistan's National Day on 23 March 2002 the first nuclear signal emanating from Islamabad, but was viewed in New Delhi as a reversal of his 12 January promises tinged with a warning to India of an 'unforgettable lesson' if it dared to challenge Pakistan. The unforgettable lesson was seen as alluding to the use of Pakistani nuclear weapons to counter an Indian conventional attack across the LC to a target audience of its domestic population, India and especially the US along with the international community. Although there was no official response to this nuclear signal by the Indian Cabinet, Defence Minister George Fernandes dismissed Musharraf's statement as 'childish' and while addressing troops on the India–Pakistan border, he asked, 'Islamabad not to mistake New Delhi's patience as weakness and force India into an armed conflict.' He suggested that Musharraf was living in a make-believe world and asked him not to underestimate India's power and potential and thus forget Pakistan's standing vis-à-vis India.[8]

The second nuclear signal from Islamabad came at a time of relative calm along the India–Pakistan borders. On 5 April 2002, German weekly newsmagazine *Der Spiegel* published the text of an interview with Musharraf, which, loosely translated, reads as follows, 'Using nuclear weapons would only be a last resort for us. We are negotiating responsibly. And I am optimistic and confident that we can defend ourselves using conventional weapons...only if there is a threat of Pakistan being wiped off the map, and then the pressure from my countrymen to use this option would be too great.'[9] The intent was probably to draw the attention of the international community to the 'nuclear flashpoint' narrative. Amidst much sensational international press coverage, the following day, a spokesman of the Pakistani government clarified that Musharraf had actually said that the use of nuclear weapons is only a last resort, if all of Pakistan was threatened to disappear from the map.

---

[8] 'Musharraf Threat Childish: Fernandes', *The Times of India*, 26 March 2002.

[9] https://www.theguardian.com/world/2002/apr/06/pakistan.rorymccarthy. This is a compilation based on Chaudhury, Rahul Roy. 'Nuclear Doctrine, Declaratory Policy and Escalation Control'. http://www.stimson.org/newpubs.cfn. https://www.stimson.org/content/nuclear-doctrine-declaratory-policy-and-escalation-control

Prime Minister Vajpayee publicly declined to comment on the Musharraf interview, stating that he 'will not like to comment till I see the entire statement'. Vajpayee never did respond to the interview.

Though India's policy to avoid nuclear signalling was maintained, there were two exceptions. The first was on 11 January 2002, a day prior to Musharraf's well-publicised address during which he promised to act against terrorists. At a press conference on 11 January 2002, the Indian Army Chief General S Padmanabhan pointed out that the possibility of a nuclear exchange between India and Pakistan was in the realm of the unknown, and that India had already declared that it would not be the first to use nuclear weapons. However, in response to a query from a journalist, the General stated that India possessed the capability of a retaliatory strike and warned that if any country was 'mad enough' to initiate a nuclear strike against India, then the perpetrator of that particular outrage shall be punished severely.[10] Although it was not intended as a nuclear threat, it was projected as one by the media. However, being contradictory to India's unstated policy on nuclear signalling, the Ministry of Defence (MoD) within hours disavowed the statement with Fernandes pointing out in a written statement that the nuclear issue should not be handled in a cavalier manner.

The second exception closely followed the first on 25 January 2002 when India flight-tested its nuclear capable medium-range Agni ballistic missile on the eve of its Republic Day. In the spirit of the Lahore Memorandum of Understanding (MoU, 21 February 1999), Pakistan was notified in advance of the impending missile test (along with the P-5 states). The official spokesperson of the Indian Ministry of External Affairs (MEA) specifically stated that, 'This (test) is not directed against any country,' though publicity given to the range of the missile—700 km—indicated the nuclear missile was

[10] *The Hindustan Times*, 12 January 2002.

Pakistan-specific. Pakistan did not respond immediately through similar tests, but did so in May 2002 by testing three of their missiles.

Phase II: 14 May 2002-17 June 2002

During this period, tensions deepened as a consequence of the 14 May terrorist attack at Kaluchak and India probably once again came close to using force. The war signals from India were at their peak, with New Delhi seemingly hard-pressed to continue threatening the use of force, whilst deliberately ignoring Pakistan's nuclear signalling.

Most leaders called for decisive action after the Kaluchak attack, but Vajpayee gave mixed signals on India's intentions. On 17 May 2002, the Indian Parliament passed a unanimous resolution condemning the Kaluchak attack and pledged to end the senseless acts of terrorism. General Padmanabhan was quoted as saying that though it was a political decision, the time for action had come.[11] Union Home Minister LK Advani on 20 May 2002, said the government would go ahead and win the battle.[12] However, on 21 May in Jammu, Vajpayee stated that he did not see any war clouds.[13] In Kupwara (Jammu and Kashmir), the following day, addressing army personnel (the author was part of the audience, being in command of a Corps Reserve Brigade), he contradicted his earlier statement by emphatically asserting that the time had come for a decisive battle and we would have a certain victory in this battle. But in Srinagar the next day, questioned on his statement on war clouds, Vajpayee stated that 'the sky may be clear, but sometimes even when the sky is clear there is lightning, but he hoped that lightning would not strike.' In a formal statement issued on the occasion, Vajpayee was quoted as having stated that India was preparing for a

[11] Chengappa, Raj and Gupta, Shishir. 'The Mood to Hit Back', *India Today*, 27 May 2002.

[12] https://www.economist.com/node/1145838

[13] http://news.bbc.co.uk/2/hi/south_asia/2003615.stm

decisive victory.[14] These contradictory and deliberately ambiguous statements certainly increased the level of uncertainty and tension.

These were, arguably, the most important though confusing and apparently conflicting Indian pronouncements at a critical juncture of the crisis. It is important to note the contents of Vajpayee's comments about a decisive battle on 22 May. While the apparent intent behind Vajpayee's speech essentially appeared to be to boost the morale of Indian armed forces personnel and provide some direction to the increasing confusion over the future course of action vis-à-vis Pakistan, the subtext was to make an impact on Islamabad and Washington, especially the latter, by indicating the limits to India's patience over Pakistan's intransigence.

This was reflected a few days later as well. On 26 May, a day before Musharraf's well-publicised second address to the nation, Vajpayee gave a stern warning to Pakistan, while at the same time, stressing the critical role the international community could play in reining in Pakistan, and averting a war. From the hill station of Manali, where he had gone on a holiday after his visit to Jammu and Kashmir, Vajpayee reflected that we should have given a fitting reply the day they attacked Parliament. Although this was subsequently clarified as not stating that we should have struck, but that it would have been better to have taken action immediately after 13 December, its import was clear.[15] Vajpayee also stated that 'world leaders told India to keep patient while condemning the 13 December attack, but India wouldn't follow the same advice now and the world should understand there is a limit to India's patience'.[16]

[14] 'Vajpayee ready for political solution to J&K issue'. 23 May 2002, http://in.rediff.com/news/2002/may/23jk1.htm

[15] http://meaindia.nic.in/mediainteraction/2002/05/28mi01.htm. Z Davis, The India–Pakistan Military Standoff: Crisis and Escalation in South Asia, Macmillian, USA, p 148.

[16] Bedi, Rahul. 'A Strike Launched', *Frontline*, 8-21 June 2002. http://www.frontline.in/static/html/fl1912/19120130.htm

Partly in response to the war rhetoric emanating from New Delhi, a senior member of the Pakistani cabinet, Lt General Javed Ashraf Qazi, told the Islamic Republic News Agency (IRNA) in Islamabad on 22 May that Pakistan would not hesitate to use nuclear weapons if its survival was at stake. As minister for railways, and a former chief of the ISI (1993-95), Qazi stated, 'If it ever comes to the annihilation of Pakistan, then what is this damned nuclear option for, we will use (it) against the enemy.' He further added, 'If Indians will destroy most of us, we, too, will annihilate parts of the adversary. If Pakistan is being destroyed through conventional means, we will destroy them by using the nuclear option, as they say, if I am going down the ditch, I will also take my enemy with me.'[17]

A week later, Pakistan's Permanent Representative to the UN in New York, Munir Akram, asserted his country's right to use nuclear weapons against India's conventional superiority. At a press conference in New York on 29 May, his second day in office, Akram stated, 'India should not have the license to kill with the use of conventional weapons while our hands are tied with regard to other means to defend ourselves. We do not wish to expend our limited resources on building up a conventional defence which will completely debilitate our development...we have to rely on our own means to deter Indian aggression. We have that means and we will not neutralise it by any doctrine of NFU.'[18]

Although none of these Pakistani statements was ever denied or alleged to have been misquoted by the media, additional pronouncements were made to alleviate their impact, in view of the possible negative international public option. In an interview to the *Washington Post* published on 26 May, Musharraf attempted to

---

[17] Varma, KJM. 'Pak Will Not Hesitate to Use Nukes Against India'. 22 May 2002, http://www.rediff.com/news/2002/may/22war10.htm

[18] Shourie, Dharam. 'Defiant Pakistan Threatens to Use Nukes'. 30 May 2002, http://in.rediff.com/news/2002/may/30war2.htm

downplay the threat of nuclear war. On being asked to describe the circumstances in which he would consider using nuclear weapons if war were to erupt, he said, 'It is such a question which I wouldn't like to even imagine, frankly, that we come to a stage where this is due. But let me give an assessment that this stage will never come. We have forces. They follow a strategy of deterrence. And we are very capable of deterring them. I really don't think we will ever reach that stage, and I only hope and pray that we will never reach that stage. It's too unthinkable.'[19]

Nonetheless, in the midst of this rhetoric, Pakistan flight-tested three types of nuclear-capable ballistic missiles. Although Islamabad also unilaterally provided New Delhi (along with the P-5 and other neighbouring states) with advanced notification of these tests in the spirit of the Lahore MoU, their timing could not be overlooked. On 25 May, the North Korean-based Ghauri (Hatf-V—1,500-km range) medium-range surface-to-surface ballistic missile was tested, followed by the Chinese Ghaznavi (Hatf-III) (DF-11) short-range (300-km range) the next day. Two days later, coinciding with the visit of British Foreign Secretary Jack Straw to Islamabad, Pakistan launched the Abdali short-range (180 km) ballistic missile. Taking place as they did, amidst the presence of some 5,000 American military personnel in Pakistan, deployed in view of the war on terror in Afghanistan, these tests also sent a message of independent military action.

Although Pakistan's nuclear signalling public statements and missile tests were viewed as extremely provocative by New Delhi, there was no reaction to them in kind, so as to not invoke Kashmir as a nuclear flashpoint. New Delhi, therefore, tried to react nonchalantly to these developments and publicly scoffed at Pakistan's ballistic missile flight tests. Vajpayee, in Manali at the time of the

---

[19] Coll, Steve. 'Pakistan Says Raids in Kashmir Have Ended; Musharraf Demands Response From India', *The Washington Post*, 26 May 2002.

first test, appeared singularly unimpressed, dismissing it as routine, and saying that India was not taking it seriously. At various times, the Indian MEA's responses to the missile tests were that it was not greatly impressed, as there was nothing indigenous about them (they were based on either imported technology or acquired hardware); moreover, they were a glaring example of missile proliferation and the tests were conducted primarily for the benefit of Pakistan's domestic audience.

On 27 May 2002, Musharraf, during a televised address to the nation, stated, 'We do not want war. But if war is thrust upon us, we would respond with full might, and give a befitting reply'. In a press conference on 28 May 2002, India's External Affairs Minister Jaswant Singh stated 'Musharraf's television address on 27 May is both disappointing and dangerous. Disappointing, as it merely repeats some earlier assurances which remain unfulfilled till today and dangerous because through belligerent posturing, tension has been added to, not reduced.'[20] In an answer to a question on the possibility of a nuclear conflict, he stated, 'No, we are not talking of it… Musharraf and some of his ministers were speaking very casually about nuclearisation.' He stated that this was tantamount to nuclearisation of terrorism, adding that 'in this we see an example of how the promotion of terrorism and the threat of nuclear weapons are being held simultaneously. The international community has to take note of the seriousness of these two dangers'[21]. A few days later, Fernandes, participating in the International Institute for Strategic Studies (IISS) Shangri-La Dialogue in Singapore, queried, why 'world opinion is not reacting to such open threats of Pakistan on use of

[20]See transcript of press conference by Jaswant Singh, External Affairs Minister on 28 May 2002. www.pmg.org.za/docs/2002/minutes/020612pcforeign.htm

[21]Transcript of press conference by Jaswant Singh, External Affairs Minister, on 28 May 2002. Op cit.

nuclear weapons. Is this not an attempt to blackmail India and the rest of the global community?'.[22]

Vajpayee went even further, a few months later. Addressing the fifty-seventh session of the United Nations General Assembly in New York on 13 September 2002, Vajpayee stated, 'Dark threats were held out that actions by India to stamp out cross-border terrorism could provoke a nuclear war. Nuclear blackmail had emerged as a new arrow in the quiver of state-sponsored terrorism.' He went on to say, that to succumb to such blatant nuclear terrorism would mean forgetting the bitter lessons of the 11 September tragedy.[23]

On 2 June 2002, India's Defence Secretary Yogendra Narain told the New Delhi-based weekly newsmagazine *Outlook* that India would retaliate with nuclear weapons if Pakistan used its atomic arsenal; both countries were, therefore, required to be prepared for mutual destruction.[24] This statement, like the earlier statement by General Padmanabhan, was promptly denied as it contradicted India's policy on nuclear signalling. The press release from the MoD itself stated that the government made it clear that India did not believe in the use of nuclear weapons and neither did it visualise that it would be used by any other country.[25]

Lessons of Nuclear Signalling

The Indian emphasis during the crisis and during both the phases was to downplay the nuclear threat so as to weaken the narrative of an India–Pakistan nuclear flashpoint. During the first phase

[22] Address by George Fernandes at IISS Asia Security conference, Shangri-La Dialogue, Singapore, 2 June 2002. https://www.stimson.org/content/nuclear-doctrine-declaratory-policy-and-escalation-control

[23] PM's address to the fifty-seventh session of the United Nations General Assembly. http://www.un.org/webcast/ga/57/statements/020913indiaE.htm

[24] 'India Will Use Nuclear Weapons If Pakistan Does: Defence Official', *The Hindustan Times*, 3 June 2002.

[25] 'War If At All, Will Be Sans Nukes', *The Pioneer*, 4 June 2002.

(13 December 2001-13 May 2002), India categorically threatened the use of conventional force against Pakistan. In the second phase (14 May 2002-17 June 2002), an added Indian dimension was the appeal to the international community to rein in Pakistan's support of terrorism. During both these crises phases, India avoided any nuclear signalling and ignored Islamabad's nuclear signalling.

In periods of heightened military tensions, public statements, as a method of conveying messages and creating fear in the minds of opponents, were unpredictable in terms of their impact, as interpretations varied among different types of audiences and mostly exacerbated the prevailing atmosphere of ambiguity and confusion.

Some lessons emerge from this narrative on India–Pakistan nuclear signalling.[26]

- Signalling is a two-way street. A signal may not always be perceived as intended, since it may be missed or distorted. Confirmation or rebuttal of the signal requires acknowledgement that are monitored and relayed back. If these are missed, the signaller may perceive the other's absence of action as a response, albeit deliberately low-key, forcing it to raise the stakes in a subsequent signal, and thereby further exacerbating tensions. A case in point was Pakistan's alleged movement of nuclear-tipped ballistic missiles for deployment during the Kargil Conflict in July 1999, which New Delhi may not have been aware of, and Islamabad's consequent reactions, if the conflict had prolonged.
- A non-signal may also be perceived as a signal as is it not always clear when a specific action or an event is perceived as a signal, even though it may not actually have been intended as such. Specific actions may have technological

[26]Reproduced from Chaudhury, Rahul Roy. 'Nuclear Doctrine, Declaratory Policy and Escalation Control', op cit.

or bureaucratic dynamics, independent of ongoing political tensions. Also, it is not always the case that specific actions are always well thought out and extensively deliberated within governments prior to signalling.

- Signalling is confused by multiplicity of actors. Although New Delhi perceived signals to its multiple constituencies as fairly clear and unambiguous, they were not perceived as such, due to the number of principal actors involved. Clearly, the principal signaller was Prime Minster Vajpayee. In addition, statements of cabinet ministers like Fernandes, Jaswant Singh, LK Advani, Yashwant Sinha and General Padmanabhan and Defence Secretary Narain, were amenable to be interpreted as signalling. Some statements were promptly rebutted and indicated incoherence within the government.
- In both New Delhi and Islamabad, it was exceedingly difficult to interpret the other's signals and in reality it merely increased the confusion and uncertainty.

The Role of the United States

The US emerged as the principal mediator between India and Pakistan and led international diplomatic efforts aimed at conflict avoidance.[27] The salience of the nuclear factor and the war in Afghanistan in driving American diplomacy was indeed significant. The Taliban regime in Afghanistan had been overthrown after 9/11 and Hamid Karzai's government was sworn in on 22 December 2001. Washington was well aware of Pakistan's past involvement with the Taliban and al-Qaeda, and its role in aiding and abetting the cross-border insurgency in Kashmir. But it also mistakenly

---

[27] Powell, Colin. 'US Prevented India–Pakistan War', *The Indian Express*, New Delhi, 18 July 2004. Quoted as one of the achievements of the Bush administration at a round table for correspondents in Washington, 20 May 2004.

believed that President Musharraf's regime was perhaps the best antidote to Islamic militancy in Pakistan. The US was also keen to strengthen the hands of President Musharraf in an attempt to prevent nuclear weapons from falling into the hands of the radical anti-Western elements in Pakistan. Therefore, the US attempted to safeguard its emerging partnership with India by leaning on Pakistan to terminate support for the Kashmir insurgents in a manner that avoided humiliating President Musharraf publicly or compromising his domestic credibility. Above all, Washington sought to prevent a nuclear conflagration in the region.

The Role of Deterrence

Even if the role of deterrence in the India–Pakistan crisis of 2002 cannot be proven, some cause and effect relationships are discernible. The presence of nuclear weapons resulted in India adopting the concept of limited war and Pakistan espoused the notion that India would not respond conventionally to cross-border terrorism due to the nuclear factor. When India threatened war, Pakistan retracted, albeit temporarily. At that point, it could be said that nuclear weapons tilted the scales against war. This was indeed a significant strategic effect on the continuum of the India–Pakistan strategic process. India's limited military challenge to the India–Pakistan status quo was an indication of continued frustration, but it may have also been a signal of successful deterrence by Pakistan of more dangerous aggressive acts by India.

At the operational level, India could have hoped for some sort of victory in January 2002, since Pakistan had to move several of its formations from its western borders that were supporting the US war in Afghanistan. The presence of the US armed forces in Pakistan would have complicated military operations and involved significant negative political fallout. Conventional deterrence could have failed, but did not, most probably because of Musharraf's

12 January speech. The relative strengths of both armed forces were undoubtedly in India's favour.

Table 7.1: India/Pakistan Relative Strength: 2002

	India	Pakistan
Total Strength	1,298,000	6,20,000
Army		
Regional Commands	5	Nil
Field Armies	4	Nil
Corps	12	9
Armoured Divisions	3	2
Infantry Divisions	29	19
Independent Brigades	15	22
Navy		
Aircraft carriers	1	Nil
Submarines	16	10
Principal Surface Combatants	27	8
Air Force		
Aircraft	701	366

Source: Military Balance: IISS, London, 2002-03, pp 129, 133.

Though India had a preponderant advantage in air and naval forces and only marginally for the army, due to land border with China, arguably a swift and decisive victory was not easy to accomplish. Also, it was never clear what would victory imply, though capture of large tracts of territory and maximum attrition of Pakistan's armed forces potential, were the generally assumed objectives. By May 2002, the prospect of a decisive victory had receded further, since both armed forces had been mobilised for nearly five months and surprise was difficult. However, if the Indian operational objectives were indeed limited, the possibility of achievement endured. How limited operational objectives could realise the political objective of forcing Pakistan to stop trans-border terrorism was never clear.[28] In all probability, nuclear weapons as the elephant in the room, seems the overarching factor that prevented war.

The deterrence effect of nuclear weapons has played a major part in both the creation and management of the crises. Nuclear deterrence was manipulated by India, Pakistan and the US to serve their respective interests. Apart from averting a nuclear war, an India–Pakistan war would have been disastrous for the US war in Afghanistan, given a reprieve to elements of the Taliban and al-Qaeda and Pakistan may have utilised their services against India.

The crises were the closest both countries came to testing their respective belief systems regarding limited war in a nuclear environment. In the end, both countries claimed victory. India believed that it had succeeded in coercing Pakistan to admit to its armed support to insurgency and extracted promises to withdraw support.[29] On the other hand, Pakistan claimed that it had succeeded in deterring India from launching a war and its nuclear deterrence stand had been vindicated.[30] In hindsight, India knows that Pakistan reneged on its promises and its coercive diplomacy had only minimal and marginal success.

There were divergent voices in India about who deterred whom. The Prime Minster said that war was imminent at a point of time

---

[28] Several discussions with senior military officers regarding this issue have been held. I am convinced that there was a gap between political objectives and its realisation by military means in the crisis of 2002. At best, military means could temporarily disturb the status quo but not change it. This is borne out by the events after 2002, wherein the terrorist infrastructure in Pakistan and infiltration has continued.

[29] Fernandes, George. 'Armed Forces Mobilization Met Objectives', *The Hindu*, 21 November 2002.

[30] Gen Pervez Musharraf's speech at Attock, 'We defeated the enemy without a war', quoted in *The Hindu*, 13 December 2002.

and that his government was prepared for nuclear war were it was to come about. President APJ Kalam stated that without nuclear weapons, war would have been a certainty. The former Army Chief, VP Malik, disagreed and asserted that nuclear weapons neither eliminated nor reduced the risk of outbreak of hostilities.[31]

However, it could be argued that India's decision to adopt a wait-and-watch policy after Musharraf's 12 January and 27 May speeches was influenced by other issues too. The pressure by the US and the uncertainty of the ability of limited war to deliver were also significant factors. But at the root of these factors the nuclear issue was omnipresent. Nuclear blackmail need not be one of using nuclear weapons. It is enough to create conditions for their use, to bring about a response from major powers as adequately demonstrated in the crisis.[32]

However, General Padmanabhan dismissed the notion that it was Pakistan's nuclear capability which deterred India from going to war on those two occasions. He said, 'When we assess our adversaries, we assess all [their] capabilities. We had evaluated it [the nuclear capability] and were ready to cope with it.'[33] The statement perhaps reflected the Indian belief that a conventional war could be fought successfully under a nuclear shadow.

Gains from Manipulation

India and Pakistan claimed gains from the manipulative force-threatening exercise. India obtained a promise from Pakistan of ending cross-border terrorism and in September 2002 also obtained breathing space to proceed with the elections in Jammu and Kashmir, which it conducted very successfully. It obtained the US commitment

---

[31]Raghavan, VR. 'Manipulating Nuclear Deterrence', *The Hindu*, 3 July 2002.

[32]Ibid, Raghavan, VR.

[33]Bidwai, Praful. 'Musharraf's Speech Raises Nuclear Danger', 2 January 2003, www.antiwar.com

to keep Pakistan on track to curb support of terrorism in Jammu and Kashmir. However, for India, the reprieve was temporary, because Pakistan reneged on its commitments.

Pakistan obtained the critical advantage of India's plans of going to war being put on hold and the US emerged as a guarantor of peace between India and Pakistan.

As already noted in the earlier chapter, decision makers in India and Pakistan learnt divergent lessons from Kargil. The lessons from the 2002 crisis were not dissimilar and notably, this time too, the lessons were divergent, buttressing the challenges of future showdowns.

LESSONS LEARNT BY INDIA

- The war scare succeeded in placing terrorism in South Asia in the same context as the war against the al-Qaeda network.
- The limitations of coercive diplomacy need acknowledgement and its effectiveness requires that the threat of coercion is viable. It also becomes counterproductive if it is extended beyond a particular time frame. Therefore, India's mobilisation cannot be repeated over and over again.
- Avoid being placed in a position where neither de-escalation nor punishment of Pakistan seems possible.
- Brinkmanship could be a viable policy up to a limited extent and can also have significant negative economic consequences like travel warnings and evacuation of non-essential personnel, as the international community is prone to overreaction.
- The US intervention in the last two crises has benefited India. It has not only accepted the US 'facilitation', but had manoeuvred the US into being the guarantor of Pakistani promises. Of course, in hindsight this too proved transitory.

Lessons Learnt by Pakistan

- The combination of conventional and nuclear deterrence works. By sending strong messages and by mobilising its forces, Pakistan believes it succeeded in calling India's bluff. Because deterrence works, Pakistan need not make fundamental changes to suit India, or succumb to pressure from India.
- India's coercive diplomacy was based on an erroneous assumption and miscalculation that Pakistan was isolated and economically weak and would not be able to sustain the pressure that India could apply on Islamabad. It also misread the reaction of the international community, which did not pressurise Pakistan as much as it had hoped for.
- Pakistan's position on Kashmir may be eroding internationally, especially in the US. However, Washington still needs it for its continued operations in Afghanistan and Washington cannot be entirely in India's camp. Pakistan's support to jihadis going across the LC has to be modulated, but Pakistan need not give up its Kashmir position in return for nothing.

Implications for the Region

Although the crisis had waned temporarily, it could return with a vengeance in the course of time. Indian and Pakistani leaders appear to have learnt lessons that brinkmanship and deterrence work; that the US can successively intervene. India is pleased that coercive diplomacy has succeeded, while Pakistan is confident that it has withstood India's threat.

The LC has been ever more sanctified, as cross-border infiltration is no longer tolerable to the US and the international community and raises the possibility of converting the LC to an international border as a pragmatic solution to the Kashmir issue.

Even if the role of deterrence in the India–Pakistan crisis of 2002 cannot be proven, some cause and effect relationships are discernible.

The presence of nuclear weapons resulted in India adopting the concept of limited war and Pakistan embracing the notion that India will not respond conventionally to cross-border terrorism. India's limited military challenge to the India–Pakistan status quo was an indication of continued frustration. But it may also be an indication of successful deterrence by Pakistan of more dangerous aggressive acts by India. In hindsight, it is apparent that Musharraf's 12 January and 27 May speeches were tactical retreats to tide over the crisis.

At the operational level, India could have hoped for some sort of quick victory in January 2002. Conventional deterrence, therefore, could have failed but did not, most probably because of Musharraf's speech. But in May 2002, while the prospect of an Indian victory had receded, the possibility of war existed if the Indian operational objectives were indeed limited. In any case, the nuclear cloud could not be wished away and in many ways seemed to have fulfilled its role—prevention of large-scale conflict. In both instances, the role of nuclear deterrence, amongst other factors, seems the predominant reason that prevented war; since the primary driver of American diplomacy as well as India and Pakistan's political decision-making, was the nuclear factor.

Despite divergent lessons learnt by India and Pakistan, easing of political and military tensions resulted in the India–Pakistan Ceasefire Agreement of November 2003. However, Pakistan continued to send infiltrators to fight its proxy war in Jammu and Kashmir. Pakistan's duplicitous involvement in the Afghan War also continued to deepen. In retrospect, it should have, therefore, been no surprise that the use of terrorism as Pakistan's primary tool for foreign policy played out in a major terrorist attack in Mumbai in November 2008 thus creating another India–Pakistan crisis. Unlike in the 2001-02 crisis, India did not react militarily, but instead preferred to respond on political and diplomatic levels. The question that needs to be answered is why India eschewed a military response?

CHAPTER 8

DETERRENCE REVISITED—MUMBAI TO URI

MUMBAI ATTACKS (2008)

Strategic Context

Consequent to the defusing of the 2001-02 crisis, the India–Pakistan relationship showed a positive trend. Both countries signed a ceasefire agreement in November 2003 that applied to the LC, as well as the international border and Siachen Glacier. The ceasefire agreement endured until the first major violation on 13 May 2008 when Pakistani troops opened fire in the Tangdhar Sector in Jammu and Kashmir. The major political-level initiative was the institution of a composite dialogue. The decision to hold a composite dialogue followed a meeting in Islamabad on the sidelines of the South Asian Association for Regional Cooperation (SAARC) summit between India's Prime Minister Atal Bihari Vajpayee and Pakistan's President Pervez Musharraf in January 2004. The first round of the composite dialogue at the Foreign Minister's level was held in September 2004. The composite dialogue encompassed a fairly large basket of issues:

confidence-building measures, Jammu and Kashmir, Siachen, Sir Creek, Wullar Barrage, terrorism, drug trafficking, economic and commercial cooperation and promotion of friendly exchanges in various fields.

Pakistan's internal security situation had deteriorated substantially consequent to the Pakistani Army assault on Lal Masjid in July 2007, with the intent to clear it of armed militants. By the end of 2007, LeT operations were well underway in Afghanistan as well and it was openly claimed by a senior member in December 2007 who noted that the group was mainly focused on operations inside Afghanistan.[1] In a clear attempt to assist the Pakistan Army in its counterterrorism policy, it also adopted a policy of providing logistic, financial and training support to any jihadi group on the condition that such assistance would be directed against Afghanistan and not misused for activities inside Pakistan. Along with these developments, inputs indicate the JuD/LeT greatly increased its presence across Pakistan. Violence spread into the Pakistani heartland of Punjab and by the end of 2008, the total casualties due to terrorist violence were more than the previous three years combined.[2]

By August 2008, nearly 100,000 Pakistani troops were deployed on its western borders. Also, the spring and summer of 2008 witnessed the highest coalition causalities since the beginning of the war. General Ashfaq Parvez Kayani, the Pakistan Army chief, was under pressure from the US to 'do more for Afghanistan' and neutralise not only the al-Qaeda, but also the Tehrik-e-Taliban Pakistan (TTP) and LeT elements who were aiding the war in Afghanistan.[3] This was conveyed in a meeting held on 28 August 2008 aboard the *USS Lincoln*.

---

[1] http://www.satp.org/satporgtp/sair/Archives/sair11/11_44.htm

[2] This was estimated conservatively at 7,997 killed and 9,670 injured. Pakistan Institute of Peace Studies, Pakistan Security Report, 2008.

[3] https://www.dawn.com/news/956860

The India–Pakistan composite dialogue commenced in April 2003 and continued in fits and starts despite several terrorist attacks, till the Mumbai attack of December 2008. In July 2006, the serial train blasts in Mumbai killed 187 and injured 824 persons and though the dialogue was suspended, it was resumed after a meeting on the margins of the Non-Aligned Movement (NAM) Summit in Cuba between India's Prime Minister Manmohan Singh and Pakistan's President Musharraf in September 2006. In February 2007, it was also revealed that India and Pakistan had, through 'back channels', agreed to a non-territorial solution. The solution also included free trade and movement across the LC. However, further progress could not be made as Musharraf ran into trouble with Pakistan's internal politics,[4] which also had stalled the fourth round of talks that was to be held in October 2007. On 13 May 2008, nine serial blasts in Jaipur killed seventy and injured 186 people. However, since the perpetrators were unknown,[5] the fourth round, which had already been scheduled for 21 May 2008, was held between India's Foreign Minister Pranab Mukherjee and his Pakistani counterpart Shah Mehmood Qureshi.

The lowest point in the 2004-08 period of the composite dialogue process was the suicide attack on the Indian embassy in Kabul on 7 July 2008. It killed fifty-eight people, mostly locals. Among the dead were India's Defence Attaché Brigadier Ravi Mehta and a diplomat VV Rao. Indian and the US intelligence indicated that Pakistan's ISI had masterminded the attack though Pakistan denied it as usual.

---

[4] Parashar, Sachin. 'Manmohan Singh, Musharraf came close to striking Kashmir Deal: Wikileaks', *The Times of India*, 3 September 2011. https://timesofindia.indiatimes.com/india/Manmohan-Singh-Musharraf-came-close-to-striking-Kashmir-deal-WikiLeaks/articleshow/9841701.cms

[5] The Rajasthan police had later arrested fourteen persons who allegedly belonged to the Students Islamic Movement of India (SIMI). In 2011, eleven of the fourteen were acquitted by the court.

The Indian Foreign Secretary Shivshankar Menon acknowledged that the peace process was under stress.[6]

The Kabul suicide attack was followed by several terrorist attacks between July and October 2008; Bangalore (25 July), Ahmedabad (26 July); New Delhi (13 September); and Guwahati (30 October). Though commonly attributed to being part of Pakistan's proxy war, there was not much evidence that was produced regarding their culpability. Overall, the atmosphere of India–Pakistan peace was in any case vitiated by the time the Mumbai attacks took place on 26 November 2008. In Pakistan, the political situation after the General Elections (February 2008) had brought in a coalition of the Pakistan People's Party, the Pakistan Muslim League (Nawaz) and the Awami National Party to power. This unlikely coalition showed little ability to govern, but appeared to desire an improvement in relations with India. This was more apparent after the election of Asif Ali Zardari as President (6 September 2008) when in his first address to the Parliament, he declared new initiatives to start cross-LC trade, more liberal visa regimes, and full support to Kashmiris for their 'fundamental rights' rather than self-determination. The establishment, however, remained wary of the new President. Even more seriously, the President, in an interview to the *Wall Street Journal* (4 October 2008), compared jihadi groups fighting in Kashmir to terrorists and 'wolf packs' who became more active when India and Pakistan were talking peace. The interview also quoted Zardari as saying, 'India has never been a threat to Pakistan' and called for strong economic ties with India.[7] The statements led to protests in Baramulla and Anantnag in Jammu and Kashmir, causing Information Minister Sherry Rehman to issue a disclaimer and reiterating Pakistan's continued diplomatic support for the Kashmiris. Alongside these events, however, levels of violence were

---

6 https://in.reuters.com/article/idINIndia-34610320080721

7 Audio clips online, https://www.wsj.com/articles/SB122307507392703831

steadily reducing in Jammu and Kashmir, and the 2008 figures were less than half for the earlier years. With the President calling for greater cooperative measures, it appeared the Jammu and Kashmir factor was, at the very least, headed for some accommodation and lessening of tensions. From the point of view of the Pakistani Army, however, the most damaging statement with regard to India came from the President in a satellite interview to the *Hindustan Times* Leadership Summit (22 November 2008) where he affirmed that Pakistan would never use nuclear weapons first against India.

The Attack

On 26 November 2008, ten terrorists launched an assault on multiple targets in India's commercial capital of Mumbai. The attacks lasted sixty hours till 29 November, leaving 166 people dead and 304 wounded and included twenty-five foreigners killed and twenty-one injured of different nationalities. The terrorists, it was later learnt, had been launched from Pakistan's port city of Karachi on 21 November and had landed at a beach in Mumbai at around 8-8:30 pm on 26 November. They split into five two-man teams and attacked their targets with AK-47 rifles and hand grenades. Among the many targets attacked, the Taj Mahal and Oberoi–Trident hotels; Mumbai's biggest railway station, the perennially crowded Chhatrapati Shivaji Railway Terminus (CST); Nariman House, a Jewish centre; and Café Leopold, a popular hangout for foreigners and locals, were the primary ones.

India's Reaction

The fact that the Mumbai attacks involved ten Pakistani terrorists who had been launched by a boat from Karachi was known to India's leadership sometime during the night of 26 November 2008 as transmissions of mobile conversations were being monitored by Indian intelligence agencies. The initial focus was more on ascertaining lapses in intelligence and security, especially coastal security. The CCS met on 29 November to discuss India's response, the discussions about military options were inconclusive.[8]

During a meeting convened by the defence minister after the CCS meeting, the issue of escalation was flagged by the service chiefs and the government was advised to be prepared for escalation once the military options were exercised.[9] The armed forces could not provide the degree of assurance that there were reasonable chances of keeping the military action in the realm of operations other than war. The political leadership was not convinced that the use of the military instrument could achieve the political objective of getting Pakistan to abjure the use of terrorism as a tool of foreign policy. A restricted use of the military, in the form of selective strikes on known terrorist camps could, on the other hand, prevent the Pakistani civilian political leadership from taking action against the jihadi forces which they assured they would do. The Pakistani assurance eventually turned out to be a charade and the previous experiences, especially of the 2001-02 crisis and Kargil, should have provided sufficient grounds for scepticism. In the end, during the period of the crisis that stretched till January 2009 and though military options were examined by the political leadership, being unconvinced of its utility, approval was denied. A close watch was kept on military movements by the National Security Advisor MK Narayanan so as to prevent the setting-off of an action–reaction cycle with the Pakistanis.

On 2 December 2008, India's External Affairs Minister Pranab Mukherjee was quoted as saying, 'Nobody is talking of military action.'[10]

---

[8] Samanta, Pranab Dhal. '26/11: How India debated a war with Pakistan that November', *The Indian Express*, 26 November 2010. https://timesofindia.indiatimes.com/india/Manmohan-Singh-Musharraf-came-close-to-striking-Kashmir-deal-WikiLeaks/articleshow/9841701.cms

[9] Ibid.

[10] http://www.theguardian.com/world/2008/dec/02/mumbai-attack-india-pakistan

But consequent to a blame and denial cycle that played out, the mention of military options surfaced from time to time.[11] In the end, India adopted only political and diplomatic measures to get Pakistan to act against the planners and perpetrators of the Mumbai attack. Pakistan took some half-baked measures and has generally continued to preserve the LeT and other terrorist groups as its strategic assets.

India's Response

There was ample evidence that the Mumbai attack was launched by the LeT based in Pakistan. Pakistan's initial and subsequent reaction was of denial. However, as India produced evidence, they adopted the 'non-state actors beyond official control' stance and continue to do so. On 7 December, as pressure mounted from India and the US, Pakistan carried out a few arrests including those of the LeT operational commander Zaki-ur-Rehman Lakhvi and Zarrar Shah, both of whom had personally controlled the operation through mobile and satellite phones. Pakistan acted against Hafiz Saeed, the LeT chief, only after the UN ordered sanctions against the Jamaat-ud-Dawa (JuD). The JuD was a known clone of the LeT and was established as a front for it. Saeed was put under house arrest till June 2009 after which he was freed by the Lahore High Court, citing lack of evidence.

The question that begs an answer is why did India confine itself to political and diplomatic measures even though it had ample proof of the Mumbai attack being launched from Pakistan by the LeT? The reaction was in marked contrast to India's reaction in 2001. Several reasons could be construed:

---

[11] On 19 December, Pranab Mukherjee, who had earlier ruled out military action, stated that 'India would consider the entire range of options' in order to 'protect our people from this menace'. https://www.hindustantimes.com/india/india-obliged-to-consider-entire-range-of-options/story-uAw7hTVeBzswlXLnRSzABI.html

- Though Pakistan's initial reaction was one of denial, it immediately condemned the attack and offered to cooperate in the investigation. Zardari's offer to send the DG ISI, though shot down by the Army Chief Kayani, reflected the schism between the civilian leadership and the Pakistani army. This raised the possibility of the Mumbai attack being launched with the support of the ISI and therefore, the Pakistani Army without the concurrence of the civilian leadership. In the weeks following the attack, Pakistan took at least limited action to arrest the perpetrators and raised some hopes in India that the civilian leadership was committed to bringing the perpetrators to justice and taking action against terrorist groups.
- Military options provided by the Indian military leadership did not convince the political leadership that they would resolve the issue of preventing terrorism originating from Pakistan. The military options of air strikes and ground action could result in a limited war but how that war could facilitate the political objective of getting Pakistan to eschew using terrorism against India as a foreign policy tool was neither clear nor convincing enough to the political leadership. Apart from not achieving the main political objective, considering its focus on economic development, a limited war was not in India's interest.
- Immediately after the Mumbai attack, Pakistan made it clear that if India was to react militarily it would have to shift most of its troops from the western border, thus weakening its effort against the US-led Afghanistan campaign against the Taliban. This threat was aimed to get the US to put pressure on India not to react militarily. The US, in fact, did send its representatives, including Secretary of State Condoleezza Rice, to counsel restraint. India, on the other

hand, perceived the possibility that the strategic purpose of the Mumbai attack was a result of a conspiracy between the ISI, LeT, the Taliban and al-Qaeda to get Pakistan to ease pressure on their bases at the Af–Pak border. An Indian military response could mean playing to the terrorists' game plan.

Internalised Narratives

As had happened with the 2001-02 crisis and the Kargil Conflict, the Mumbai attack produced divergent internalised narratives on both sides. The divergent narratives provide fertile ground for future misunderstanding and miscalculation.

India's views on Pakistan's duplicity were reinforced and the inadequacy of political and diplomatic measures to force Pakistan to give up the terrorist tool as a foreign policy instrument were also internalised. However, the political leadership showed very little inclination to reach for the military instrument. Some attribute this to the personalities involved in the decision-making, especially the Prime Minister who continued to keep faith in seeking a stable and friendly relationship as part of the nation's grand strategic vision. Though calls for military action again filled the Indian public discourse, the fact that the Congress government was voted back to power in 2009 indicates that India staying with political and diplomatic measures did not have any major impact on its electoral fortunes nor did it find much ballast in the electoral debate. The moot question remains: how will India react to another major terrorist attack that originates from Pakistan and can it gamble on Indian military inaction the next time around?

India's future reaction to another major terrorist attack is impossible to guess. But two factors could tip the scale towards military action. The first will depend on the make-up of the political leadership. A right-wing polity could see their electoral future jeopardised by military inaction and therefore, be more inclined to military action. Secondly, the Indian media could arouse passions that will hinge on jingoism and force the political hand towards military action. Revenge would be the emotional driver, and especially, when it comes to India–Pakistan issues, its inflammable potential is well known on both sides.

India's military inaction has strengthened Pakistan's internalised narrative that India will not react militarily due to Pakistan's threat for escalation to the nuclear realm. Military minds on both sides are burning brainpower to achieve operational solutions that they hope will overcome the structural constraints imposed by the nuclear shadow. The clash of belief systems in the India–Pakistan deterrence dynamics is pregnant with mistrust, misunderstanding and miscalculation and more so during a crisis. The danger is rooted in Pakistan's miscalculation of India's reaction and India reacting emotionally to violence inflicted on its society. The urge for revenge could cloud reasoning, which is a phenomenon that has littered the landscape of human conflicts. It is difficult to make a case of the India–Pakistan situation escaping history's natural orbit unless both countries are guided by political sagacity that escapes the irresistible force of passion grounded in a sense of revenge. The moot question is will the fear of nuclear use keep the uneasy peace?

DETERRENCE PERSPECTIVES: 2014-17

The assumption that for the most part of the UPA tenure, India–Pakistan relations had improved, received a hard knock after several terrorist attacks of varying intensities occurred in the Indian hinterland during the period of UPA government's second tenure (2009-14)—Pune (13 February 2010), Varanasi (7 December 2010), Mumbai (13 July 2011), Delhi (7 September 2011), Hyderabad (21 February 2013) and Patna (27 October 2013). None of these attacks, which were mainly attributed to home-grown terrorist groups, primarily the Indian Mujahideen, were viewed as convincing enough to elicit a military response.

The BJP had fought the national elections of May 2014 on a plank that projected an embrace of a tough stand against cross-border terrorism and painted the previous government as being weak and soft in dealing with terrorism and Pakistan. After assuming power in May 2014, consecutive terrorist attacks brought public pressure to fulfil their promise, with the first challenge to the government's tough stance on terrorism presenting itself in the east, on the Indo-Myanmar border in April–June 2015.

Indo-Myanmar Border Attacks

A string of attacks in the states of Arunachal Pradesh (Tirap-2 April 2015), Nagaland (Mon-3 May 2015) and Manipur (Chandel-4 June 2015) carried out on army convoys by insurgent groups using Myanmar as their base, were suspected to have been undertaken primarily by the NSCN (K), an insurgent group that had withdrawn from an earlier ceasefire agreement. The fatalities in Arunachal, Nagaland and Manipur were three, eight and eighteen soldiers. Within four days of the Manipur attack, on 8 June 2015, India carried out raids with its special forces on two insurgent camps in Myanmar. The official statement at the conclusion of the raid stated, 'The army engaged two separate groups of insurgents along the Indo-Myanmar border at two locations. Significant causalities have been inflicted on them. As a consequence, threats to our civil population and security forces were averted. While ensuring peace and tranquillity along the border and in border-states, any threat to our security, safety and national integrity will meet a fitting response.' Although the official statement never acknowledged that raids were carried out in Myanmar, India's Minister of State for Information and Broadcasting, Rajyavardhan Singh Rathore stated, 'It was a much-needed decision that was taken by our Prime Minister. This decision was extremely bold in nature. It involved our special forces crossing the border and going deep into another country…. We will not tolerate any strikes on India or Indians. We will always wield the initiative on either being friendly or engaging in aggressive action. We will strike at a place and time of our choosing.' When asked about the possibility of similar strikes on Pakistan based terror groups, Rathore, elaborated, 'This is a message for all countries, including Pakistan and groups harbouring terror intent towards India. A terrorist is a terrorist and has no other identity. We will strike when we want to.'[12]

In hindsight, and from a deterrence perspective, the Myanmar raids neither deterred the NSCN (K) or other insurgent groups from carrying out further attacks nor did it force the Myanmar government to state unequivocally, that their territory is not used as safe bases. It did indeed have a temporary impact that lasted for a year as the period June 2015-June 2016 hardly witnessed any attack by insurgent groups, presumably due to a tactical pause when the insurgent groups reassessed and readjusted their deployments, bases and plans. Commencing July 2016, the insurgent groups became active again and resumed their activities. On 10 July 2016, the Divisional Commissioner's office in Manipur's Senapati district was damaged in a bomb explosion and was followed on 28 August 2016 by an ambush on the Assam Rifles in the Changlang district of Arunachal Pradesh. On 16 September, a four-vehicle convoy of Assam Rifles was ambushed in Nagaland's Zunheboto district. Any further raids by India was also circumscribed by the Myanmar's government's frosty reaction to the June 2015 raids that required diplomatic energy to assuage the violation of sovereignty protests made by the political leadership of Myanmar.

[12] The Anatomy of Army's Surgical Strikes on militants in Myanmar, *Hindustan Times*, 10 June 2015.

Arguably, the raids provided only temporary relief perhaps because it required a sustained retaliation policy which was politically problematic to undertake. As the raids were projected as symbolic of the tough stand adopted, the major gains from the raids were in the realm of domestic politics. The point that was sought to be conveyed was, 'This strong and tough political leadership will not hesitate to use force to protect our national interest'. Such a narrative blended seamlessly with the prevailing popular sentiment that demanded a muscular stance in dealing with terrorism.

Rathore's statement elicited responses from Pakistan's political and military leadership. Interior Minister Nisar Ali Khan stated: 'Pakistan is not like Myanmar' and at a conference of formation commanders, Army Chief General Rahel Sharif 'took serious note of the recent hostile rhetoric coupled with covert and overt actions to destabilise Pakistan'.[13]

Following the ascent to power of the Modi government in May 2014, India had extended an olive branch to Pakistan which unfortunately had a short shelf life due to the repeated reassertion of a familiar mode of derailment through cross-border terrorist attacks. Earlier, the Lahore Bus Yatra by Vajpayee was followed by Kargil in 1999; Manmohan–Musharraf talks by Mumbai serial train bombings and later by Mumbai 2008. Similarly, the terrorist attack at Gurdaspur on 27 July 2015 was preceded by a Modi–Sharif meeting at Ufa on 10 July 2015 where they reached an understanding to resume the composite dialogue; subsequently Modi's surprise visit to Lahore on 25 December 2015 for a meeting with Sharif was followed by Pathankot on 1-3 January 2016.

Gurdaspur/Pathankot Terrorist Attacks

Both the Gurdaspur and Pathankot attacks occurred at times when the civilian leadership on both sides were attempting to restore

[13] Ibid.

political relations. In both cases, it seems, officially, there was no serious thought to any Indian reaction that involved the use of cross-border application of force. Pakistan claimed innocence as usual and offered their sympathies. After Pathankot, Pakistan agreed to assist in the investigation and in a surprising move their investigating team was given access to the scene of action inside the Pathankot Air Base and also to areas en route the base from the infiltration site on the border. After the investigators returned to Pakistan, they went back to the default mode—denial, deceit and obfuscation. A follow-up visit to Pakistan by Indian investigators never materialised. These experiences therefore hardened the Indian stand that settled at 'no talks without halting terrorism'. The Uri attack in September 2016 coupled with loss of twenty soldiers therefore presented no political choice but to do something that may deter Pakistan from further attacks, assuage domestic anger and preserve the tough image of the government.

Uri–September 2016

In the early hours of 18 September 2016, four Pakistani terrorists who had infiltrated through the highly guarded Line of Control in Kashmir, attacked an administrative camp of the Indian Army which was located in the border town of Uri, claiming the lives of eighteen soldiers and injuring several others. The attack was carried out at a time when India–Pakistan relations were at a low ebb due to the Gurdaspur and Pathankot attacks. In addition, Kashmir was witnessing a prolonged phase of violent civil protest that followed in the wake of the killing of a local terrorist leader of south Kashmir, Burhan Wani. On the international canvas, the United Nations General Assembly (UNGA) was in session from 13 to 26 September.

The culpability of Pakistan was established through electronic tracking especially the GPS carried by the terrorists, intercepted communications and other items of equipment and food items found on the terrorists. Following a familiar pattern, Pakistan denied culpability, floated conspiracy theories and accused India of trying to distract attention from human rights excesses in Kashmir. Also, another phase of cross-border ceasefire violations commenced with fairly extensive use of mortar and artillery fire. All this was equally predictable and followed the existing pattern. But what was different this time was the internal political dynamics in India.

The substantial loss of lives coupled with the deterioration in India–Pakistan relations presented a challenge to the Indian government that was left with no choice but to pick up the gauntlet. The Indian reaction was announced on 29 September post the UNGA Session. The announcement was made through a press conference specially called for the purpose. The Indian Director General of Military Operations (DGMO) initially alluded to Pakistan's continued culpability in cross-border terrorist activities. The Indian reaction was described as: 'Based on receiving specific and credible inputs that some terrorist teams had positioned themselves at launch pads along the line of control to carry out infiltration and conduct terrorist strikes inside Jammu and Kashmir and in various metros in states, the Indian Army conducted surgical strikes at several of these launch pads to pre-empt infiltration by terrorists. The operations were focused on ensuring that these terrorists do not succeed in their design to cause destruction and endanger the lives of our citizens. During these terrorist operations, significant casualties were caused to terrorists and those providing support to them. The operations aimed at neutralising terrorists have since ceased. We do not have any plans for further continuation. However, the Indian armed forces are fully prepared for any contingency that may arise. I have been in touch with the Pakistan Army DGMO and informed him of our actions. It is India's intention to maintain peace and tranquillity in the region. But we cannot allow the terrorists to operate with impunity and attack citizens of our country at will.

In line with Pakistan's commitment in January 2004 not to allow its soil or territory under its control to be used for attacks against India, we expect the Pakistani army to cooperate with us to erase the menace of terrorism from the region.'[14]

Indeed the scope and effect of the surgical strike was a remarkable achievement by India's special forces especially because considerable damage was inflicted without suffering any casualties.

The Indian DGMO's statement evoked considerable jubilation amongst the Indian public and the government projected the 'surgical strike' as a feather in its political cap. In the realm of domestic politics, 'surgical strike' became symbolic of a government that meant business and one that would not hesitate to take action in the interest of the nation. The term 'surgical strike' was added to the lexicon of domestic politics and later even used to describe the surprise sprung on the Indian public by a demonetisation exercise aimed against black money. 'Surgical strike' thereafter figured in all the elections held at the state levels and was touted as a major achievement of the government. To a public that was hungry for action against Pakistan, 'surgical strike' was food for revenge. The electoral benefits for the ruling dispensation also seemed considerable. From a deterrence perspective, 'surgical strike' had a confused canvas because Pakistan denied that any such strike had taken place!

Pakistan's political and military leadership completely denied India's claims of having carried out military actions across the Line of Control and the Indian government also decided against providing any proof of the strike; it, therefore, relieved the Pakistanis from undertaking any military reaction.

The lessons from a deterrence perspective are once again in the eyes of the beholder and differed for both sides.

[14] Press Statement by DGMO, Press Information Bureau, Government of India, Ministry of Defence, 29 December 2016.

India

- Surgical strikes signalled a switch in Indian attitude regarding the use of military power to counter terrorism.
- Surgical strikes could be an effective military tool to inflict damage to Pakistan in order to deter it from conducting terror attacks.
- Pakistan's reaction to surgical strikes could be muted or be dealt with and confined to reasonable escalation limits.
- Considerable domestic political benefits could be garnered if the Indian public was exposed to the narrative of India having resorted to use of force and succeeded in causing damage to Pakistan. Action taken must be displayed for public viewing.

Pakistan

- India's response to terror strike is primarily meant for the domestic polity. Nuclear weapons would ensure that no major Indian military action ensues.
- Pakistan is capable of countering any military action taken by India.

The Uri terror attack and its aftermath, the 'surgical strike', followed a familiar pattern of military discourse that has characterised the exchange of firepower across the Line of Control for several decades. The difference, this time, was perhaps in the scope of the cross-border action and, more importantly, the publicity given to the ownership for action. It also sought to convey a new approach that could be described as 'we shall hit you harder each time'.

The Nagrota attack on the army's officers mess on 29 November, just two months after the surgical strike once again tested the Indian response. The response was immediate but contained within cross-border artillery exchanges that lasted for several weeks and eventually waned. Nothing much had changed in strategic terms but in the

Indian public imagination the government had adopted a tougher stance to deal with Pakistan.

The policy of publicising India's military reaction was again witnessed on 23 May 2017 when the Indian Army publicised its 'punitive fire assault' on a Pakistani military post as a reaction to continued cross-border infiltration and several terror attacks that included the beheading of two soldiers. A video was released of the strike that portrayed the destruction being caused by firepower. Pakistan quickly countered by also releasing a video that supposedly showed the destruction of an Indian military post. A 'theatre of the absurd' was on public display, the only difference was that instead of it being a very private viewing it was exposed to the public on both sides. The hunger for vengeance was being assuaged through graphic scenes of destruction being inflicted. Domestic impact of military force application outweighed the deterrent effect. The army was now more 'heard' than 'felt' and the successes were claimed by the ruling party as its own.

The exchange of firepower across the Line of Control provides perhaps the fastest and easiest means of inflicting damage. It was also not uncommon that it endangered the civilian population who lived in the vicinity and often forced their evacuation. Overall, as long as both sides have a free will, firepower exchanges could continue ad infinitum without giving any relative advantage to either side. The ceasefire agreement of 2003 was an acknowledgement of this reality. But, after the Uri attack, the ceasefire agreement was de facto moribund.

In summary, nothing had changed except new names were added to the lexicon of discourse in India–Pakistan relationship and a new imagination of 'India acting tough' seemed to have gained ground in the realm of domestic politics. From a deterrence perspective, the tone of the military activities had heightened in tandem with an increase in Pakistan's cross-border terrorism. Terrorism and cross-border energy exchanges of firepower were re-established as the lingua franca. It was back to the situation prevailing prior to the ceasefire in 2003 except that the internal situation in Kashmir had morphed into a combination of terrorism and intensified civil unrest. This conflation carries within it the potential to activate the nuclear shadow that is shaped by the theory of limited war.

CHAPTER 9

INDIA–PAKISTAN LIMITED WAR

The political object—the original motives for war—will, thus, determine both the military objective to be reached and the amount of effort it requires.

—**Carl von Clausewitz**

Clausewitz on the Concept of War

In 1832, a year after his death, when Carl von Clausewitz's widow published his book, *On War*,[1] it was and is still acknowledged as 'the profound, comprehensive and systematic examination of war'.[2] Clausewitz earns the considerable attention he receives primarily because his theoretical writings on war and strategy yield the most powerful of conceptual tools available to understand the course and outcome of conflict. Although a theoretical work, *On War* has been considered to be of immense practical value for policymakers, strategists and military commanders at the highest level. Although the

---

[1] Clausewitz, Carl von. *On War*, Howard, Michael and Paret, Peter, eds and trans, Princeton University Press, New Jersey, 1976.

[2] Ibid, Brodie, Bernard. 'The Continuing Relevance of "On War"', quoting Herbert Rosinski, p 45.

relevance of *On War* to contemporary conflicts of the period has been questioned during the industrial age mass warfare and the nuclear era, Clausewitz, even if he has been subjected to different interpretations and selectively exploited to prove a point, has endured. Clausewitz warns very early in his treatise 'that in war more than in any other subject, we must begin by looking at the nature of the whole; for here more than elsewhere the part and the whole must always be thought of together'.[3] For, unlike the natural sciences, different variables or factors cannot be isolated and studied independently. Parts, therefore, should only be studied in the context of the whole. Limited war should be viewed in the context of the entire spectrum of conflict.

Clausewitz argues that given the human nature, war cannot be studied as though it is an exact science. In fact, he states that war is neither an art nor a science but an act of 'human intercourse', and falls in the realm of what is known today as social science.[4] Therefore, though war can be studied systematically, it is ultimately an art that requires creative solutions in specific or unique situations.

Clausewitz defines war as 'an act of force to compel our enemy to do our will'.[5] The definition implies the survival rather than the destruction of the enemy. Also, what distinguishes war from any other activity is the use of force and bloodshed. This definition must be read alongside another definition: 'War is a clash between major interests, which is resolved by bloodshed—that is the only way in which it differs from other conflicts.'[6] In addition, Clausewitz offers yet another definition: 'Essentially, war is fighting, for fighting is the only effective principle in the manifold activities generated as war.'[7]

---

[3] Clausewitz, Carl von. *On War*, op civ t, p 75.

[4] Ibid, pp 148-49.

[5] Ibid, p 75.

[6] Ibid, p 149.

[7] Ibid, p 127.

Clausewitz not only sees war as inevitable but also as a common and legitimate instrument that states must sometimes use to protect their vital interests.

Clausewitz distinguishes between war in theory and war in practice. By asking why war in practice differs from war in theory, he develops important ideas about war. The method leads Clausewitz to develop concepts, such as, friction and uncertainty in war; the rationale, that is, the political direction of war; the difference between offence and defence; total and limited war.

Clausewitz's description of war in theory identifies three inherent types of interaction in war that in theory and oftentimes in practice lead to escalation to the extreme. These, he says, are the maximum use of force (physical force), the aim to disarm the enemy (the objective of war) and maximum exertion of strength (moral forces). However, while evolving what he terms 'real war', Clausewitz explains why the extreme nature of war in theory is moderated in reality by factors such as political (rational) calculations; the inability to use all forces at once; the difference in strength between the offence and defence, insufficient or inaccurate intelligence on the relative strength of the opponent; and aversion to risk or other psychological considerations.

Clausewitz introduces a comprehensive framework for the study of war[8] and argues that war always includes rational and non-rational elements, material and moral factors, planning, control, as well as uncertainty, friction and chance. Such a framework has endured because all of these complementary and, at times seemingly, contradictory elements deal with every dimension of warfare. The framework is summarised in his famous 'trinity',[9] which is 'composed of primordial violence, hatred and enmity, which are to be regarded as a blind natural force; of the play of chance and probability

[8] Ibid, p 77.

[9] Ibid, p 89.

within which the creative spirit is free to roam; and of its element of subordination, as an instrument of policy, which makes it subject to reason alone'. Primordial violence, hatred and enmity derive themselves from the people; chance and probability relate to the armed forces; policy subordination belongs exclusively to the government. These can form innumerable combinations reflecting the character of each war. Throughout, Clausewitz fully recognises the importance of non-rational (as well as rational) factors such as charisma and creativity of the leader; the morale and motivation of the people; the influence of danger and pressure in war; the effect of uncertainty, friction, chance, uncorroborated information and intelligence; and on the ability to make rational and rapid judgements.

In Clausewitz's theory of war, the maximum use of force is not based on rational calculations but on the inherent dynamic nature of interaction. Clausewitz compares the nature of war with a chameleon. This comparison is akin to Sun Tzu's comparison of war to water: 'And as water has no constant form, there are in war no constant conditions.'[10] What clearly emerges is that there are limits to the rational conduct of war and the leadership should be aware of it before embarking on war.

Clausewitz confines absolute war to the realm of theory and, therefore, what finally defines the use of force is not the inherent dynamics of war, but the determination of the political authorities. This resolve decides the objective of war and what achieving a given objective is worth in terms of resources to be invested and which in turn determines how much counterforce the enemy will have to employ. Accordingly, war is not one uncontrolled clash of all forces available; instead it is a calculated political decision that can range from the extreme use of force to minor engagements. The analysis of war in reality clearly implies the existence and logic of limited war.

---

[10]Tzu, Sun. *The Art of War*, Sawyer, Ralph D, trans, Boulder, Colorado, 1994, p 101.

Clausewitz stresses that attention must be paid to the tension between the desire to wage war as rationally as possible as against viewing it as a carefully calculated affair with the added need to recognise the limits on rational calculations.[11] There are, thus, inherent trends in war towards the extreme (absolute war) and the moderating influence of rational political calculations on limited war. The issues are shown in the model:

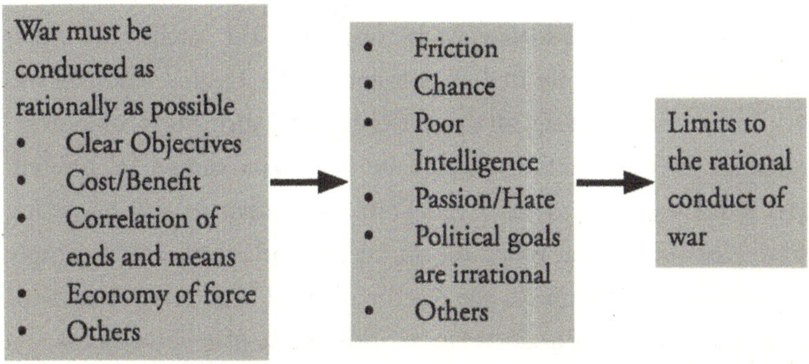

Source: Michael Handel, *Who Is Afraid of Carl von Clausewitz?* US War College Press, New Port, 1997, p9.

Clausewitz identifies a complete spectrum in warfare that spans the extremities of violence from the maximal to insignificant levels with the extent of violence being moderated by political aims and the means considered worth the sacrifice. It was this Clausewitzian legacy of the association between political aim and means that created the dilemma for force application in the nuclear era.

Limited War: Nuclear Era

Western strategists resurrected limited war as a solution to problems that originated with the advent of nuclear weapons. With nuclear weapons, it was no longer necessary to first destroy the adversary's armed forces to impose one's will on the adversary of which the

[11] Clausewitz, Carl von. op cit, p 579.

Japanese capitulation after Hiroshima and Nagasaki best illustrates the point. The situation became somewhat intractable when both belligerents possessed nuclear weapons. It became obvious that failure of deterrence between nuclear-armed adversaries could result in mutual suicide. Clausewitz's absolute war seemed possible though it implied loss of rationality and the abdication of political guidance. Thus, the end state was not what the adversaries desired and, therefore, they should have every impulse to avoid it.

Captain Basil Liddell Hart is considered the intellectual father of the nuclear era limited war theories. To Liddell Hart it was axiomatic that because of the suffering and disruption of normal relations between nations caused by war, wars should be limited. His achievement was to develop an alternative to the prevailing strategies of total war based on sound principles rather than simply an emotional response to the horror of it all. His critique of total war predated the atomic weapon. But, soon after Hiroshima, he wrote that any unlimited war waged with atomic power would be worse than nonsense—it would be mutually suicidal.[12] Strategists became preoccupied with two related problems: how to fight a limited conventional war and also wield nuclear threats without unleashing an uncontrolled nuclear war. The present India–Pakistan predicament is not dissimilar.

In 1954, Bernard Brodie argued that the availability of the atomic weapon as a deterrent will be increasingly limited to only the most outrageous kind of direct aggression. He felt that it was necessary to explore ways of consciously limiting those conflicts, which one may be unable to avoid.[13]

In 1957, Robert Osgood stressed that the principal justification of limited war lies in the fact that it maximises the opportunities for the effective use of military force as a rational instrument of policy.

[12] Liddle Hart, BH. *The Revolution in Warfare*, Faber & Faber, London, 1946, pp l-2.

[13] Brodie, Bernard. 'Unlimited Weapons and Limited War', *The Reporter*, 18 November 1954.

But the strategy of deterrence, that underpinned limited war, had to be credible and, which in turn, required that the means of deterrence be proportionate to the objectives at stake.[14]

The resurrection of limited war theories emanated from the failure of the American strategy of massive retaliation. America's possession of nuclear weapons did not avert the Korean War, their defeat in Vietnam, influence the course of the Suez crisis or the Soviet–Egyptian arms deal.

The American theory of limited war in the nuclear period was developed to explain what occurred in Korea as a source of possible precedent for future conflicts.[15] The core inspiration was a limited conventional war as in Korea, but the theory ventured to the zones of tactical nuclear war and even limited strategic nuclear war. The idea was to achieve the utilisation of force in the nuclear era without suicidal consequences. These American strategists took comfort from Clausewitz's real or limited war and wrote as if the idea of limited war would be a panacea for all strategic ills.[16] But whatever were the technical solutions that were evolved, all of them relied on the highly unlikely condition of mutual trust; both sides had to be convinced that the other was playing by the rules for it to work. Also in strategies like 'counterforce, flexible response, assured destruction', it would be impossible to tell if this was the case until the missiles had hit the cities and not the military targets that the rules dictated they were supposed to strike. However, it was argued that it was better to have a series of credible conventional and nuclear

---

[14] Osgood, Robert. *Limited War: The Challenge to American Strategy*, University of Chicago Press, Chicago, 1957, p 26.

[15] Kaufmann, William W. 'Limited Warfare', in Kaufmann, *Military Policy and National Security*, Princeton University Press, Princeton, New Jersey, 1956, pp 102-36.

[16] Osgood, Robert. *Limited War: The Challenge to American Strategy*, University of Chicago Press, Chicago, 1957; Brodie, Bernard. *Strategy in the Missile Age*, Princeton University Press, Princeton, New Jersey, 1959, Ch. 9; Halperin, Morton H. *Limited War in the Nuclear Age*, John Wiley and Sons, New York, 1963.

options rather than just one that would necessitate an act of collective suicide.

The Cuban Missile Crisis of October 1962 cast all these problems into sharp relief. Political and military realities in terms of faulty interpretation, misconception and deficient information made nuclear war much less controllable than theorists like Thomas Schelling, Herman Kahn and others had claimed. The Vietnam strategy evolved by Robert McNamara was essentially a conventional war version of the counterforce/no cities theory: using force as an instrument of coercion and withholding a larger force that could kill the hostages of the enemy's cities if it didn't back down.[17] The 'Rolling Thunder' campaign of 1965-68 was the most concerted coercive campaign, a massive strategic bombing campaign that was designed to illustrate American firepower and tenacity while also conveying to North Vietnam the costs of continued support to the Viet Cong. The idea was to gradually increase the pressure, but never to commit the full force of American power—to coerce leaders rather than defeat the army on the battlefield. The campaign failed. The carefully calibrated (if not carefully executed) attacks on civilians and economic targets around Hanoi came to be perceived as immoral in terms of human rights, as well as ineffective, encouraging the anti-war movement and tainting the reputation of those scholars who had devised the strategy.[18]

Limited Objectives

Limited war in the nuclear age was designed to achieve limited objectives. Brodie writes that if progress is to be made towards resolving the military dilemma, the one basic proposition which must be established in the minds of men is this: 'Limited war must

[17] Kaplan, Fred. *The Wizards of Armageddon*, Simon & Schuster, New York, 1983, p 329.

[18] Pape, Robert. *Bombing to Win: Air Power and Coercion in War*, Cornell University Press, Ithaca, New York, 1996, p 3.

also mean limited objectives.' Kissinger described it as: 'It reflects an attempt to affect the opponents will but not to crush it; to make the conditions imposed seem more attractive than continued resistance to strive for specific goal and not for complete annihilation.'[19] The objectives, therefore, were conceived with limitations and were expected to contain the conflict within desired levels.

Defining limited war, Osgood stated, 'Limited war is one in which the belligerents restrict the purposes for which they fight to concrete, well-defined objectives that do not demand the utmost military effort of which the belligerents are capable and that can be accommodated in a negotiated settlement.'[20] There is, of course, a clear-cut assumption of symmetry between the two belligerents in both capabilities and intentions. One could argue that the question of intentional limitation of war arose only when there was a reasonable symmetry of capabilities, sufficient to ensure that the outcome of a conflict was not a foregone conclusion.

It was undeniable that limited objectives were a necessary condition for limited war. But that still did not guarantee a limited war. Limited war was not a theory about the primacy of political objectives over military means, but of the primacy of military realities over political objectives.

Role of Friction and Chance

Military realities under the nuclear shadow implied that uncertainty resulting from friction and chance created a potential for collective suicide by loss of political guidance in the application of military force. Friction being an inalienable product inherent to any clash of arms primarily resulting from the inability to perceive reality due to the chaos and utter confusion that is the normal ambience of warfare. Western theories on limited war were criticised for their assumption

[19] Ibid, p 140.

[20] Osgood, Robert. op cit, p 18.

that a reliable grammar of limited war could be applied that which would both dominate the logic of policy and tame many of the problems of friction, chance and uncertainty.[21] They seemed to elevate technique and operational virtuosity over politics.[22] The theories required a cooperative foe. Since rules in war are not often observed, limits in limited war had to be affirmed, and if possible, safeguarded. Diverse polities, distinctive personalities, different motivations to fight, as well as different feasible styles of war, frequently threatened to make a mockery of limited war. Vietnam demonstrated that the American theory was flawed. For, it was realised that it was one thing to wage war in a limited way in support of limited objectives; it is quite another to wage such war to a successful outcome.[23]

By the 1970s, both American and Soviet planners conceded the need for extreme care. The prospective dynamics of conflict rendered any resort to nuclear use, a gambler's throw; Clausewitz's caution seemed relevant–'In the whole range of human activities, war most closely resembles a game of cards.'[24] More presciently, Clausewitz's insistence that it is the nature of war to tend towards an absolute expression of violence presaged perfectly the fear over unwanted escalation that came to dominate strategic theory and practice in the 1960s and 1970s.

The American theorists of limited war had little to offer on the topics of fanatically committed enemies, incompetent people or societies whose people were unwilling to accept sacrifices in rationally limited wars for limited purposes. But it cannot be disputed that it was necessary to confront the challenge of the time—how can force be used without suicidal consequences? The debate continued and

[21] Clausewitz, Carl von. *On War*, op cit, pp 85-86, 101-02, 119-21, 140. Clausewitz describes them as an unholy trinity that can prove lethally disabling.

[22] Gray, Colin S. op cit, p 335.

[23] Ibid, p 336.

[24] Clausewitz, Carl von. op cit, p 86.

has not yet been sanctified with an acceptable theory. Surely, the fundamental problems that plagued the theory of limited war earlier also continue to haunt India's limited war assumptions in the India–Pakistan context. The arguments eventually hinged on the issue of escalation and Clausewitz's warning that violence between adversaries will tilt to the extreme unless it could be guided by political rationality and which continues to haunt strategists looking for a solution for force application in the nuclear era. The danger of escalation was and still remains the wild elephant that defies being tamed.

Escalation

The potential of escalation did also undermine the theory of limited war. Its proponents, however, argued that it was conceivable that the adversary could be persuaded that it would not be possible to achieve his objectives with the means being applied. It was, therefore, believed that the adversary could be convinced that he could not achieve his objectives without expanding the scope of the war in terms of time and resources. Consequently, it could be demonstrated that the costs of fighting outweighed the perceived benefits. The advantages of terminating the conflict were greater than the advantages of continuing it.[25]

Clausewitz has noted the most important factor in modifying the absolute nature of war. The inherent theoretical tendency to escalate to the extreme is reined in by political calculations that introduce the rational calculations of ends and means, costs and benefits. But he warns that danger, uncertainty and chance, which he classified as friction, would make achievement of rationality a challenge.[26]

---

[25] Kaufmann, William W. op cit, p 113.

[26] Clausewitz, Carl von. op cit, p 122. His summation on the topic of friction in Book 1 is revealing: 'We have identified the danger, physical exertion, intelligence and friction as the elements that coalesce to form the atmosphere of war, and turn it into a medium that impedes activity. In their restrictive effects they can be grouped into a single concept of friction.'

A central thesis of *On War* is that war is an instrument of policy with policy being the inherently intelligent faculty that guides the instrument of war. It would be a horrendous mistake if war were to establish its own purpose. Of course, the policy cannot ask the armed forces to do what they are not designed to accomplish. As policymakers cannot micromanage the details of military operations, it is the policy, therefore, that determines the goals for which war is fought, influences the conduct of war and crafts the very definition of victory and defeat.

Even though nuclear weapons pose severe challenges to the conduct of successful war, war is always a possibility and, therefore, it is argued, that means must be devised to stop war at the lowest possible level where victory could be claimed. In his seminal work, *On Escalation*, Herman Kahn first developed the concept of escalation ladders, as an analyst's tool, which he intended would stretch the mind and stir the imagination in examining the course of potential conflicts—from low-level crises to all-out wars. Kahn's treatment of the abstract conflict spectrum led to the development of a generalised escalation ladder with forty-four rungs, commencing with an emerging crisis (in which political, economic, or diplomatic gestures are the appropriate responses) all the way up to insensate (unlimited, sustained) war.[27]

It is believed that it is possible to control conflict and determine outcomes even if it was not going according to preconceived plans. This was thought possible due to what came to be called as 'escalation dominance'. It is explained as the capacity to enable the side possessing it to enjoy marked advantages in a given region of the escalation ladder, with the proviso of other things being equal. Escalation dominance is considered a function of where one is

[27] Kahn, Herman. *On Escalation: Metaphors and Scenarios*, Penguin Books, Baltimore, 1968, p 39.

positioned on an 'escalator ladder'. It was thought to depend on the net effect of the competing capabilities on the rung being occupied, the estimate by each side of the consequence if the confrontation moved to other rungs and the means each side were equipped with to shift the confrontation to other rungs. It was one variable affecting each side's relative fear of eruption. The side which has the least to lose by eruption, or fears eruption[28] less, will automatically have an element of escalation dominance.[29] This, however, meant that opposing sides could compare their fears in the midst of chaos and confusion and be possessed of 'escalation dominance', which is impractical, especially during a nuclear exchange.

It was argued that deterrence between nuclear adversaries is far more effective in regulating conflict because the prospect of nuclear war was terrifying enough to frighten policymakers into behaving more or less rationally in crisis situations. Thomas Schelling portrayed the threshold to war as generally ambiguous and difficult to define. Escalation is characterised by uncertainty as neither side can really know what action on its part will trigger war. It, therefore, becomes an exercise in competitive risk-taking with diplomatic victory going to the side willing to take the greatest risks, assuming that war is avoided. Adversaries are expected to avoid war by taking small, carefully controlled and, whenever possible, reversible steps up the rungs of the escalation ladder.[30] Game theories and mathematical modelling are increasingly used to illustrate the potential of coercion and brinkmanship. These are scientific delights that completely ignored political ground realities, friction, and importantly, that

---

[28] Herman Kahn equated the term 'eruption' to Morton H Halperin's use of 'explosion' to describe escalation in *Limited War in the Nuclear Age*, John Wiley and Sons, New York, 1963, p 3. Explosion was described as the sudden transformation of a local war into a central war by the unleashing of strategic nuclear forces. The term 'expansion' on the other hand was a gradual increase in the level of military force employed.

[29] Ibid, p 290.

[30] Schelling, Thomas C. *Arms and Influence*, Yale University Press, New Haven, 1966, p 96.

adversaries were not strategic animals but ordinary mortals, who, when confronted with delicate and difficult situations, could be prone to innumerable fragilities that could lead to unimaginable consequences.

Another side of the nuclear coin, is the 'provocation threshold' which means that there exists an increase in the number of actions that come within the purview of coercive bargaining, but which in the nuclear age can precipitate war.[31] There was an expansion of the crisis escalation ladder which somewhat contradicted the assertion that nuclear adversaries behave cautiously because of their fear of war. Paradoxically, nuclear weapons induced caution in the adversaries and, therefore, raised their toleration limits before resorting to acts that seriously risk nuclear war. The obvious danger was that one side or the other, convinced that the step it was about to take would be acceptable to the adversary, and perhaps was likely to compel him to concede, may push the confrontation beyond the point of no return.

The possibility of miscalculations by policymakers in a crisis situation always exist. Richard Ned Lebow in his study of crises and wars states that 'policymakers are prone to distort reality in accordance with their needs even in situations that appeared relatively unambiguous'.[32] The more numerous and ambiguous the signals, the easier it becomes to do this, for 'uncertainty is a breeding ground, not of restraint but of irrational confidence'.[33] As most situations are likely to be usually fraught with uncertainty, wishful thinking arising from it cannot be precluded.

Violence, especially war, is a confused, highly unpredictable and uncertain activity, depending on decisions made by fallible

---

[31] Kahn, Herman. op cit, pp 96-102.

[32] Lebow, Richard Ned. *Between Peace and War*, John Hopkins University Press, Baltimore, 1981, p 277.

[33] Ibid, p 277.

human beings organised into imperfect governments, depending on fallible communications and warning systems and on the untested performance of people and equipment. It is, moreover, a hot-headed activity in which commitments and reputations can develop a momentum of their own.[34]

Manipulation of Risk

International relations often acquire a character that can be described as competition in risk-taking, which involves not tests of force but that of will, best amplified by the India–Pakistan crisis of 2002. The question that arises involves a degree of risk that is worth taking. There is also the aspect of the evaluation of the risk involved in a course of action. The anxiety from uncertainty and the sheer unpredictability of catastrophic events blur things, change their character and add an entire dimension to military relations: The manipulation of risk.[35]

In the nuclear era, uncertainty will be the enduring feature of war. Errors of judgement regarding the opponent's reactions and intent are inevitable. Random events and false alarms will be endemic, leading to some decisive action to hedge against the unforeseeable.[36]

Coercion through nuclear threats has been the methodology practised both during the Cold War and in India–Pakistan conflicts.[37] Coercion has been differentiated into two types—deterrence and compellence. Deterrence means stopping an undesired action from occurring (Pakistan forcing India not to launch a limited war). Compellence is reversing an undesired action that has already taken place (India forcing Pakistan to stop supporting terrorism in Jammu

---

[34] Schelling, Thomas C. *Arms and Influence*, Yale University Press, 1966, p 93.

[35] Ibid, p 94.

[36] Ibid, p 95.

[37] Coercion according to Thomas Schelling is most successful when threats need not be even carried out. See, Schelling, Thomas C. *Arms and Influence*, op cit, p 71. See also, Ellesberg, Daniel. 'The theory and practice of blackmail', in Oran Young, ed, *Bargaining: Formal Theories of Negotiation*, University of Illinois Press, Urbana, 1975, p 344.

and Kashmir). The creation of risk, usually a shared risk, is the technique of compellence and is best described as brinkmanship. It is a competition in risk-taking which initiates a process that carries some risk of unintended disaster.[38] In practice, however, it is difficult to always distinguish deterrence and compellence since coercion is a dynamic process as adversaries normally counter-coerce the coercer.

Brinkmanship came to be described as 'manipulating the shared risks of war' that meant exploiting the danger that someone might inadvertently go over the brink, dragging the other along. But what was important in brinkmanship was the requirement of the element of uncertainty or anticipated irrationality. The risk of disaster was the most potent element in the situation that could be manipulated to exploit and intimidate.

Limited War as Strategy

Limited war as a deterrent or as a compellent means of intimidation could be viewed as an action that enhances the risk of a greater war. The occurrence of a limited war increases the probability of a greater war. There are, therefore, two dimensions to the threat to engage in limited war. One in which the threat is to directly inflict costs in terms of casualties, expenditures, loss of territory, loss of face or anything else, while the other is the threat to expose the adversary and oneself to a heightened risk of larger war.

The engagement in limited war is akin to starting to rock the boat.[39] It sets in motion a process that is not altogether in one's control.

---

[38]For the difference between deterrence and compellence, see Schelling, Thomas C. *Arms and Influence*, op cit, p 70. One of the major differences he cites is that deterrence involves the setting of the stage where the act that is intrusive, hostile or provocative is usually to be deterred; the deterrent threat only changes the consequence if the act in question—the one to be deterred—is taken. Compellence in contrast usually involves initiating an action or an irrevocable commitment to action that can cease or become harmless only if the opponents respond.

[39]Ibid, p 105.

Deliberately raising the risk of greater war is, thus, a tactic that may fit the context of limited war, particularly for the side most discontent with the progress of war. Schelling believed that the threat of limited war might be potent even when there was little expectation that one could win it.[40] But the potential of the threat to spin out of control once military forces became engaged could not be predicted. Paradoxically, the same uncertainty, it was claimed, provided the means to extract gains through the manipulation of risk. It was obvious that risk should be commensurate with the value of political objectives and that no political objective could be worth a nuclear exchange. Collective suicide was, of course, sheer madness.

The Indian Viewpoint

The Lahore Declaration issued at the end of the meeting by Prime Minister Atal Bihari Vajpayee and his Pakistani counterpart Nawaz Sharif in early 1999 affirmed that 'the nuclear dimension of the security of the two countries adds to their responsibility for avoidance of conflict between the two countries'.[41] But a few months later, Pakistan's betrayal at Kargil, forced a rethinking in India and resulted in the resurrection of an Indian version of the theory of limited war. Essentially, the strategy of limited war is in the context of compellence since deterrence had already failed when Pakistan started its proxy war and continued to intensify it.

In January 2000, at a seminar held at the Institute for Defence Studies and Analyses (IDSA), Defence Minister George Fernandes and Army Chief General VP Malik expressed official Indian views on limited war. In his opening address, the defence minister explained the rationale of limited war:

---

[40] Ibid, p 109.

[41] Memorandum of Understanding, signed by the Indian and Pakistan Foreign Secretary, Lahore, Pakistan, 21 February 1999. https://www.stimson.org/lahore-summit

'We had understood the dynamics of limited war.... Nuclear weapons did not make war obsolete, they simply imposed another dimension on the way warfare could be conducted.... So, the issue was not that war had been made obsolete by nuclear weapons, and that covert war by proxy was the only option, but that conventional war remained feasible, though with definite limitations, if escalation across the nuclear threshold was to be avoided....'[42]

At the closing address of the seminar, General VP Malik, stated:

'The rationale for limited war is economic considerations, risk of high casualties, international pressure and the nuclear factor.... Limited war is characterised by limitations, which tend to control its conduct and space. It could be limited in time, geographical area or force level. In the Indian perspective, when we talk of a limited war, it could range from the kind we have been engaged in Siachen since the 1980s, the Kargil War in the recent past, to future wars, which could be fought by us in confined sector(s) along our long frontiers. It is not possible to outline the exact contours of any future war with great deal of certainty.... The whole scope and conduct of limited war would be governed by the end state.'

Jasjit Singh, the Director of the Indian think tank, IDSA, in an article 'Dynamics of Limited War', defined limited war as the following:

'If nuclear war and total global war are no longer viable propositions as an extension of politics by other means, the only choice available to states to use destructive forces for political purposes is through limited conventional war, sub-conventional war with military-type weapons, and the use of coercive military force without necessarily resulting in war. The overall result has been reducing the potential of war down to limited wars...'[43]

---

[42] www.idsa-india.org/newpap-sph.html

[43] Singh, Jasjit. 'Dynamics of Limited War', *Strategic Analysis*, Vol-24, No. 7, October 2000, p 1205.

Speaking at a seminar in November 2002, General VP Malik elaborated on the concept of limited war:

> 'The fundamental point for a limited war is that it is a political process conducted for bargaining. The aim is not to win but rather not to lose and fight in such a way that the enemy is forced to settle for peace.... There is a linkage between deterrence and escalation. Capability to wage a successful conventional and nuclear war (if necessary) is an important deterrent. A war may well remain limited because of a credible deterrence.... Continuous control of the escalatory ladder by the political and military leadership is important.'[44]

It must be noted that the term 'escalatory ladder' is an oxymoron, for once you climb an escalator, you cannot get off till you reach the top. Careless use of metaphoric language can provide false but dangerous comfort in the nuclear domain.

Pakistan's Viewpoint

Pakistan holds a different view and Musharraf has stated in a speech on 12 April 1999 that while the possibility of conventional war between India and Pakistan was virtually zero because of mutual vulnerability and nuclear weapons, the proxy wars were not only possible but also very likely.[45]

This clash of India–Pakistan belief systems typically reflects the stability–instability paradox.[46] Both views seek refuge in images that are perceived as individually beneficial. The main point being missed is that in the nuclear era security cannot be unilaterally sought, but is, instead, now wholly dependent on mutual cooperation, especially in the realm of escalation—a condition that would be very difficult to obtain in the real world.

---

[44] Malik, VP. 'Limited War' in Singh, Jasjit, ed, '*Air Power and Joint Operations*', Knowledge World in association with Centre for Air Power Studies, New Delhi, 2003, p 157.

[45] The Kargil Review Committee Report, op cit, p 242.

[46] The US and Soviet Union avoided war during the Cold War, while jockeying for advantage in myriad ways, including proxy wars and a succession of crises that became surrogates for direct conflict. This tense standoff in which much blood and money was expended, but without direct conflict, was described as the stability–instability paradox.

India/Pakistan Nuclear Strategy

India has declared that its nuclear capability is only to retaliate and it has also implied that its response strike could be a large-scale society destroying one. India wants Pakistan to understand that there is no scope for nuclear bargaining through limited strikes. As explained earlier in chapter five, India, has taken the position that for better self-defence, it will respond with a massive strike even if Pakistan used a low-yield weapon against an Indian target inside Pakistani territory.[47] India wants Pakistan's choices in the event of war to be limited to conventional defeat or nuclear obliteration. Thus, India envisages that a conflict fought in the conventional domain should be controlled. On the other hand, if the nuclear threshold is crossed, control is virtually jettisoned and massive retaliation takes over.

Pakistan has not accepted this two-choice option. It wants to create more options. Towards this end, it has chosen an employment strategy that is different from India. It has taken the position that it will use nuclear weapons if certain vaguely-defined red lines like territorial loss, military loss, economic strangulation and internal stability are crossed.[48] Since 2011, Pakistan has advocated the use of battlefield nuclear weapons. Pakistan has also stated that it will use nuclear weapons in a graduated manner, starting with counter-military targets.[49] Pakistan's hope is that it can limit India's retaliatory strike through the logic that even a full-weight Indian strike will not be able to prevent Pakistan from initiating retaliation. Unfortunately for

[47] The draft nuclear doctrine states, 'any nuclear attack on India and its forces shall result in punitive retaliation with nuclear weapons' (clause 2.3). The position with regard to forces has been reiterated by Defence Minister George Fernandes in 'Military Option if Diplomacy Fails', *The Hindu*, 4 January 2002.

[48] Khalid Kidwai, Director General of Pakistan's Strategic Plans Division, quoted in Cotta-Ramusino, Paolo and Martinelli, Maurizio. 'Nuclear Safety, Nuclear Stability and Nuclear Strategy in Pakistan', Concise report of a visit by Landau Network–Centro Volta, 21 January 2002.

[49] Cohen, Stephen. *The Pakistan Army*, Oxford University Press, Karachi, 1998, p 177.

Pakistan, they can never be sure that India, imbued with greater strategic depth, would never react massively. So, in essence, its first strike nuclear threat under all circumstances is incredible. But inadvertent escalation into the nuclear realm would always remain a possibility, even though it's a situation both countries would like to avoid. Caution should, therefore, guide decision makers in the nuclear context. One could, however, argue that since the military is in power in Pakistan, they have a propensity for greater risk. Pakistan, with a politically assertive army, has internalised a high degree of aggressiveness and a sense of culture-based military superiority.[50] Democratic India stands disadvantaged in a situation where risk is being manipulated. But Pakistan, unfortunately, cannot take this as a definite proclivity; it would still be, as Clausewitz had predicted, a gamble.

Analysis of Limited War

Objectives

Limited war would entail unilateral limitation of political and military objectives. Conveyance of self-imposed restrictions poses several difficulties. The adversary could perceive it as unlimited and unacceptable. If Pakistan wishes to avoid escalating a conflict, how will it cope with an outcome that is militarily and politically unfavourable. If India wishes to avoid a nuclear first strike, there would be difficulty in conveying political and military saliencies. Unlike the Kargil Conflict, operational necessity may not always permit India from indicating geographical limits to the conflict. Indicating political limits will allow Pakistan to better plan its response.

---

[50] Gauhar, Altaf. 'Four Wars, One Assumption', *The Nation*, Islamabad, 10 September 1999. Gauhar states that Pakistan launched the 1947, 1965, 1971, 1999 (Kargil) wars on the presumption that Indians are too cowardly and ill organised to offer any effective military response. See also, Schofield, Julian. 'Militarized Decision Making for War in Pakistan: 1947-71', *Armed Forces and Society*, Vol-27, No. 1, 2000.

Moreover, lack of trust is inevitably a preponderant factor in a crisis. This poses insurmountable problems to communications between adversaries. Statements can be perceived quite differently from their original intent. Limitations stated may be ignored or simply distrusted. India's political objective that gave birth to the notion of limited war was to get Pakistan to stop its proxy war in Jammu and Kashmir. A military solution could be perceived to range from invasion and land grabs to air strikes and the threat of nuclear retaliation combined with economic sanctions and political isolation. It would, therefore, be evident that the achievement of India's objective would depend on the success of these coercive instruments. Such success cannot be unilaterally assumed but would stem from an interactive and dynamic process. Pakistan's resistance and resilience would undoubtedly be a major factor. Such resistance and resilience could become greater with India's military success and could fuel escalation. Superior military force may seem necessary to achieve the Indian political objective, and the initial expectations of military and political limitations would require upward revisions. Clausewitz's warning regarding the proclivity towards the extreme would be valid and applicable. Political caution should, at least in theory, keep the scope of political objectives highly limited. The escalation dilemma would dictate the terms. Limited war must, therefore, be seen as a product of military realities over political objectives.

Policy debates often revolve around whether military action will prove to be a quagmire or not. This metaphor not only captures nicely the dangers inherent in actions that are intrinsically hard to control, but it also directs attention to what happens after the fighting begins rather than before. This is unfortunate, because the reasons whether an opponent settles or fights, how long and how hard he fights, indeed whether he can be coerced at all, will depend in large part on judgments derived from perceptions, before the first bomb is dropped.

The Problem of Escalation

The acid test of limited war is in the ability to control escalation which cannot be a unilateral process though it might well be in the interest of both. In situations where trust is absent and communication channels are closed, the ability to control escalation would be substantially eroded. During such periods of strained relations, declaratory statements lack credibility and may be clearly insufficient. Once fighting begins, bilateral diplomacy between India and Pakistan would likely be quite limited. Instead, diplomacy would be channelled through third parties, multilateral and international organisations. The fog of war, distrust and resultant friction will pose insurmountable problems to clarity of communications. It is also possible that escalation may result inadvertently from the sheer momentum of military operations. Large-scale air operations could pose a threat to or damage some major elements of Pakistan's nuclear forces. These elements could be Pakistan's nuclear delivery means (aircraft and missiles) and command and control systems. Pakistan could perceive such attacks as deliberate attempts to neutralise its nuclear arsenal and, hence, contemplate its use before losing them. Air power has been considered to be a major source of escalation in any conflict. This is due to the fact that air operations necessitate a spatial expanse that is characterised by speed of operations and vastness. The necessity to gain control of skies inevitably results in the initial air power application being directed to destroy the air power assets of the adversary. Thus, there is an inevitable escalation in the spatial extent of the conflict that is coupled with inadvertent degradation of the adversary's nuclear assets. On the other hand, in comparison to land forces, air power is ideally suited for punitive strikes since an aircraft can strike and return quickly. The escalation and de-escalation span is, therefore, negligible. Considering the superiority of the Indian Air Force over the Pakistan Air Force, there is no doubt that air power is the ideal instrument to execute punitive strikes though its impact in a larger conventional conflict could definitively be escalatory.

Questionable Assumptions

The methodology for the conduct of limited war by the political and military leadership in the India–Pakistan context remains uncertain. If the notion of limited war rests on the assumption, 'the escalation ladder would be carefully climbed in a carefully controlled ascent by both protagonists',[51] it demands cooperation, which during a crisis or conflict is invariably absent or in short supply. The notion also does not take into consideration, friction, chance and uncertainty. The situation is compounded also by a geographical proximity that makes missile and aircraft, time to target woefully inadequate in terms of warning and reaction. Both countries have limitations in surveillance, intelligence and command and control capability. Moreover, rationality may be at a premium when fear generated in a crisis impinges on the diverse decision-making structures, which in Pakistan's case is completely military influenced. The India–Pakistan nuclear equation is also characterised by limited force structures that depend on aircraft and missile capability for delivery. Both capabilities are amenable to a dual use of conventional and nuclear warheads. It would not be possible to distinguish between an aircraft carrying conventional weapons and one carrying nuclear weapons. There is also the propensity of military forces to acquire a logic of their own that is disconnected from the political objectives. There are many historical examples of militaries striking out on offensive actions unknown to their civilian superiors. Soldiers often interpret policymaker's injunctions that allow them maximum operational discretion.[52] All these issues make escalation control difficult, especially at the military level. Escalation control, therefore, requires

[51] Gen Malik in his speech at the Limited War seminar. www.idsa-india.org/newpap-sph.html

[52] Posen, Barry. *Inadvertent Escalation*, Cornell University Press, London, 1991, p 17.

that military operations be circumscribed by politically imposed restraints that are respected and understood by both sides, a condition that could be described as wishful thinking during an ongoing war.

Faith in Nuclear Deterrence

Some believe that since India and Pakistan possess nuclear weapons, both countries will never use them and Pakistan's threats are bluffs that must be exposed. Therefore, escalation, if any, will remain in the conventional realm. To them the notion of limited war is anathema and an unnecessary and self-imposed restraint.[53] This view seems to place an underserved faith in nuclear deterrence and ignores completely the fog of war and inadvertent escalation. It debunks Clausewitz's warning that there are limits to rational calculations in war because of friction and the fog of war. The greater danger is of miscalculation, miscommunication and misjudgement especially after both sides have placed their nuclear forces on alert.

Playing the Nuclear Card

In any future war, Pakistan will flash its nuclear card as early as possible in order to invite international intervention and thus seek to blunt India's conventional edge. The nuclear card, though fundamentally incredible, will have political utility and perhaps Pakistan's military leadership could even feign to acquire the madman image.[54] Even a time-bound threat to nuke New Delhi or Mumbai, unless India ceases operations, may trigger a population exodus from these cities that could completely paralyse its essential services, disrupt its road communications and result in law and order problems that can cause

[53]This view is based on interviews with some senior serving officers of the Indian armed forces who did not want to be quoted.

[54]President Richard Nixon attempted to convey an impression of a 'madman' in October 1969 to coerce North Vietnam by ordering a nuclear alert. See also, Sagan, Scott and Suri, Jeremi. 'The Madman Nuclear Alert', *International Security*, Vol-27, No. 4, the President and Fellows of Harvard College and the Massachusetts Institute of Technology, 2003, pp 150-83.

tremendous political and economic damage to India. Coupled with international intervention, there would be immense pressure on India to halt operations. India's reaction would depend on the interest at stake and the risk acceptability of the political leadership. However, one could, argue that such a situation can be avoided if the speed of India's operational success surpasses Pakistan's ability for counter action and the international community's speed of reaction—an operational problem that was sought to be resolved through the Cold Start doctrine. This not only entails major relocations and changes in the existing Indian conventional force structure but ignores the greater probability of Pakistan's countermeasures that could negate any advantage of speedy reaction by counter positioning. At a basic level, Cold Start derives its logic from the natural proclivity of all militaries to be first off the block, by launching offensives before the adversary is fully mobilised. Historically, this natural predilection has also tipped crises into war, with the First World War being the prime example. For once mobilisation is underway, counter moves are inevitable and political control is either slackened or loses its grip over military events. At a strategic level, India's Cold Start does not logically dovetail with the requirement of balanced deployment to counter two adversaries either singly or in unison. The tactical advantage sought against Pakistan is outweighed by the potential disadvantage against China, it being the stronger adversary. Cold Start is perhaps tactically logical but it is strategically questionable. Pakistan has exploited it to magnify the Indian threat and paint the India–Pakistan border as a nuclear flash point.

Conventional Space

In practice, the 'grammar' of war and strategy shows policy as what is and is not possible. Although the political dimension of strategy necessarily is logically superior to the operational military dimension, the latter field-tests the viability of the goals generated by the former. Since control of escalation requires a cooperative opponent and has linkages with actual operational plans, it was never clear as to how the Indian military planned to conduct a limited war with more than a reasonable degree of assurance that war will not spiral into the nuclear realm. Though conventional space exists, the extent of usable conventional space remains untested. But the space is certainly hemmed in and restricted by a host of uncertainties that makes a decision for war look like a desperate gamble. Though the threat of limited war by India in 2002 did force Pakistan to promise the cessation of infiltration, the promise turned out to be a tactic that was a temporary ploy to wriggle out of a difficult situation. This seriously undermines the scope of political objectives that could be achieved through threat of limited war.

While some lessons were learnt during the Kargil Conflict of 1999 and India–Pakistan crisis of 2002, it is often the case, that two adversaries may learn divergent lessons from the same conflict. Only those perceptions tend to be retained that reinforce the original trajectory of beliefs. The extent of conventional space remains in the eyes of the beholder. Perhaps none other than a field test could prove the issue of limited war, one way or the other. But as a tool of politics and even if it is clouded with uncertainty, limited war could endure. Paradoxically, it is the same uncertainty that keeps it alive.

CHAPTER 10

THE CONFLATION OF TWO REVOLUTIONS

A nation is a society united by a delusion about its ancestry and by common hatred of its neighbours.

—**Ralph Inge**

A revolution in strategic affairs is marked by the advent of a force application phenomenon, which is not easily amenable to contestation and, therefore, drastically alters the terms of engagement between adversaries. The architecture of India's geostrategic context is overshadowed by two revolutions in strategic affairs, the first of which was the advent of the missile-mounted nuclear weapon, against which there could not be any assured level of protection. The sheer magnitude of damage coupled with the speed of destruction was a new phenomenon that drastically altered the nature of strategic relations between nuclear powers. The possession of nuclear weapons induced caution about the application of force against each other and transformed the nature of political discourse during settlement of disputes. It could be assumed with some degree of finality that nuclear weapons have curbed the scope of India–Pakistan conflicts in terms of both political objectives and geographic space. The ceiling for the containment of conflict is, however, threatened by Pakistan's use of terrorism as a foreign policy tool. Terrorism, in its contemporary form, is the child of technology and its magnified effectiveness is a major by-product of the second revolution in strategic affairs—the lethality revolution.

The lethality revolution is one in which technology, specifically the explosive and information and communication technology (ICT), has empowered individuals or small groups of people to inflict deadly destruction and damage on a scale that has never been witnessed before, and which, under specific conditions, is difficult to contest. The improvised explosive device (IED) remotely detonated, the suicide bomber, the vehicle-borne IEDs and cyber-attacks by individuals or small groups of both state and non-state actors, are a few of its manifestations to which no easy countermeasure has been possible. Insurgents, terrorists, states and non-state actors have incorporated these attack techniques and leveraged them to successfully combat even the most technologically advanced armed forces in the world of which the American experience in Iraq and Afghanistan provides ample substantiation.

It must be understood, however, that successfully harnessing the lethality revolution is dependent on the degree of support from the local population, which is possible only by winning the battle of narratives and whoever succeeds through words and deeds in getting the better story across, could be expected to ultimately win. For, without local support of the population, individuals and small groups cannot remain hidden and out of harm's way from a better armed and more technologically advanced adversary. Moreover, with an increasingly interconnected world, technology provides a plethora of communication media platforms that can transmit images of death and destruction inflicted in near real time to a global audience, thus expanding the geographic space where terror can be projected. Any geography, conducive for hit-and-run attacks, is also a major determinant of the potential of the lethality revolution. Combined with the ability to sustain itself over a long period of time, one of the major strategic effects of the lethality revolution is the difficulty it poses to occupying territory and keeping it under occupation against the will of the people, especially if the people can access technology for destructive purposes.

In the India–Pakistan context, the retention of large tracts of captured territory will be challenged unless either the local population is pacified or the area is de-populated. If local support is not obtainable to a sufficient level, the use of IEDs and the suicide bombers can raise the costs of occupation to unacceptable levels. Of course, it must be accepted that an occupier can retain territory if force is used ruthlessly like Russia did in the Caucasus region, but with a rider that there would be a heavy price to pay in the long term.

A nation like Pakistan using terrorism as a tool of foreign policy can carry out spectacular attacks using small groups and the effect of which can be incredibly magnified if the media is leveraged to project the destruction. The use of suicide groups has become the preferred technique as it reduces the scope for effective protection against such attacks. Easy commercial availability of dual-use explosive materials has facilitated a host of disenchanted groups to resort to diabolical acts of violence, the counter efforts to which need to be sustained over a very long time frame to wear down the adversary and, more importantly, require the sustained support of the population on whose behalf violence is supposedly being perpetrated. Pakistan's expectations of weakening the Indian resolve by the sustained use of terrorist acts, have, instead, only served to fortify it.

The nuclear and lethality revolutions have impacted the India–Pakistan political and strategic dynamics. Pakistan has attempted to leverage both the revolutions and paradoxically become the victim of the lethality revolution, but that has not yet diminished its proclivity to use terrorism as a tool of foreign policy. As long as Pakistan does not renounce terrorism, the India–Pakistan political and strategic context remains vulnerable to eruptions of violence that, unless contained, could sever the link between force application and political rationality, especially if nuclear weapons are alerted. The greatest danger is posed by the unintended or accidental action that could trigger an action–reaction sequence that could easily overwhelm survival instincts, with its sheer speed and scale of destruction.

Political and Strategic Context

The India–Pakistan political and strategic context remains beleaguered in the mould of the classical 'security dilemma'.[1] It could be stated with some conviction that nuclear deterrence has contained conflicts within manageable limits and both nations have acted cautiously despite indulging in rhetorical threats in times of crisis or limited hostilities. More importantly, both arsenals are de-mated and there is no substantive indication that these arsenals have been deployed or been mated during any of the crises including during the Kargil Conflict. However, nuclear dangers, mostly unfounded and overblown, have been a major dynamic that has also catalysed the US to mediate in crises and draw global attention to the Kashmir issue.

Nuclear deterrence and terrorism provide the ingredients for a potentially explosive mix that fortuitously so far has been contained within bounded rationality. The India–Pakistan situation also fits the

---

[1]'A structural notion in which the self-help attempts of states to look after their security needs tend, regardless of intention, to lead to rising insecurity for others as each interprets its own measures as defensive and measures of others as potentially threatening'. Hez, J. 'Idealist International and the Security Dilemma', *World Politics*, Vol-2 (1950), Cambridge University Press, p 157.

notion of the 'security–insecurity paradox',[2] major conflicts have been avoided, even as low-level conflicts continue in the form of terrorism emanating from Pakistan and local skirmishes on the border. The theoretical framework of the 'security–insecurity paradox'[3] is inflated in the India–Pakistan case by a host of factors that do not afford much room for an optimistic future. These factors embody a deep-rooted historical animosity that could at times of crisis breach its rational boundaries:

- Major parts of both societies harbour a deep-rooted visceral hatred for each other. Pakistan's nation-building process is yet to succeed in creating an identity for itself other than the existent one of being non-Indian. Political parties on both sides leverage the hate factor for narrow domestic political purposes, rendering political rapprochement hostage to narrow domestic political perspectives.
- Both nations nurture constituencies who believe that a solution to the India–Pakistan imbroglio requires the demise of the other as a nation state. These entities in Pakistan are not only major political parties but also terrorist groups that enjoy the covert and overt support of the state.
- Pakistan has used terrorism as a derivative of nuclear deterrence to prosecute a proxy war that seeks to bleed India through a thousand cuts. India has not reciprocated through sub-conventional means and has deliberately refrained from leveraging Pakistan's multiple fault lines, notwithstanding frequent calls to do so from various segments of the Indian strategic community.

---

[2] Summarised as 'To the extent that the military balance is stable at the level of all out nuclear war, it will become less stable at lower levels of violence.' Jervis, Robert. *The Illogic of American Nuclear Strategy*, Cornell University Press, Ithaca, 1984, p 31.

[3] Snyder, Glenn H. 'The Balance of Power and the Balance of Terror,' in Seabury, Paul (Ed). *The Balance of Power* (Scranton: Chandler, 1965), pp. 185-201.

- Till the recent advent of the NDA government, India held the consistent view that a destabilised or failed Pakistan could pose a major threat to India as it would result in a population displacement with millions seeking refuge across the India–Pakistan border. This would occur primarily due to the fact that geographic imperatives of Afghanistan to the west, the mountains to the north and the Arabian Sea to the south do not offer any alternatives. Population displacement within Pakistan has been ongoing since the Soviet–Afghan War in the 1980s with the Pashtuns, Balochis and Hindus bearing the brunt. Any major movement of population across the India–Pakistan border could, in the long term, deepen the communal fault line in India and even ultimately threaten India's integrity. Therefore, it stands to reason that a stable Pakistan is in India's best interest. However, this stance could alter under the NDA government as was signalled through an allusion to Balochistan in the Prime Minister's Independence Day address to the nation from the ramparts of the Red Fort in Delhi on 15 August 2016.
- Pakistan's armed forces have acquired much of their power by capitalising on the existential threat supposedly posed by the Indian military. The coupling of the threat to India's preparations to meet the Chinese threat serves to magnify the Indian threat to Pakistan and is typically the product of the security–insecurity dilemma. The declaration of independence by Bangladesh in 1971 has also served as a plank to hoist the Indian threat. The military, through the ISI, continues to use terrorism as a tool of foreign policy and has even sold the canard to the gullible that certain sections of the ISI sometimes operate outside the ambit of the army's control.
- Contemporary Pakistan can probably be described as a

nation that spews terrorism extensively and has a distorted democratic polity with its pillars of democratic institutions emasculated by the armed forces that continue to wield actual power, especially in the field of internal and external security; structurally it has dual power centres. Internally, the army holds the state together even as it remains the architect of most of its internal fissures. Added to this is the substantial 'jihadisation' of Pakistani society which is reflected in its intolerance of minorities and displacement of a moderate version of Islamic practices by a more fundamentalist one. Several terrorist groups, most of them spawned by the ISI, have turned against the state and fight it on the twin planks of being anti-Islamic and an American stooge.

- Most elements of the Pakistani leadership have admitted that the internal security situation is now more challenging than the Indian threat. Terrorist groups like the TTP have steadily increased their attack on the security forces.[4] Nearly 60,000 troops of the regular army have been deployed in the restive provinces of Khyber Pakhtunkhwa and Balochistan.
- The feasibility of nuclear weapons falling into the hands of jihadi groups, though discounted by the Pakistani state, has continued to draw the attention of the international community and also provided the official establishment with leverage to garner financial and military hardware support from the US by posturing in a manner that broadcasts the signal–'If not us (the moderates), it will be them (jihadis)'. In a sense, Pakistan is a state that could be teetering on the perilous edge of implosion, which the world and, especially India, cannot afford. In the contemporary situation,

[4] Prime Minister Nawaz Sharif claimed at the United Nations on 26 September 2013, 'We have lost 40,000 precious lives of men, women and children, which include 8,000 defence and security personnel,' *Tribune*, Pakistan, 27 September 2013.

ironically, a substantial military defeat of Pakistan increases the possibility of jihadi forces taking control of the state.

- Pakistan's economy has been in a parlous state for several years and has been dependent upon the US to keep it from going under. Pakistan's leverage in facilitating an orderly extraction of US forces from Afghanistan has become convenient in keeping the money flowing from the US and international financial institutions like the IMF. From 2016, the China–Pakistan Economic Corridor (CPEC), a flagship project of China's One Belt One Road (OBOR) initiative which is accompanied with an estimated $50 billion investment from China is expected to boost the economy through creation of infrastructure and energy projects. Concurrently, the Trump administration has proposed a reduction in military aid from $265 billion to $100 billion for 2018.

India's Challenge

The militaries of India and Pakistan continue in the name of deterrence to equip and prepare for war. While both pursue a course to avoid war in the nuclear realm, they yet harbour notions of achieving victory in the conventional realm. Both are grappling with a situation that defies resolution through operational virtuosity. Most military plans aim to boost deterrence but end up as military flights of fancy that are unlikely to pass any test in the practical realm. Brinkmanship in a crisis under the nuclear overhang has to be managed up to a contour, short of actual conflict. The only problem is that as it is not a unilateral game and it is virtually impossible to judge the adversary's future reaction during periods of tension, as communications and activities can be easily misinterpreted or misunderstood. There is no way of prejudging the contour of conflict till one reaches the line, by which time it may be too late to pull back. Risk-taking may work, but the cost of failure could be equally intolerable. Risks, even if they can be judged to some exactitude, will have to be weighed against the stakes at issue. In all probability, the original issues at stake will be overtaken by threats of violence that pose existential risk and look pale in comparison. In any case, how does one weigh intangibles like 'honour' and 'fear' that are often prime drivers for war and its conduct?

War, in the nuclear era, as a series of engagements, is better fought in the mind space of the opposing leadership rather than battle space. However, that is a proposition that disregards human nature and the tendency of societies to wreak havoc on each other during bouts of hatred, anger and revenge that assail the mind at those junctures. Such tendencies will always lie below the surface and can today be boosted by an active media that often transgresses its limits during crises. There is, however, much to learn from historical experiences that could preserve peace even as the dangers posed by nuclear weapons threaten existence. The challenge for India, essentially, is managing the impact of the two revolutions instead of falling prey to them.

Transformation of War

War will, however, remain an enduring feature of the human condition and it is difficult to visualise any reason that war will not continue to remain a constant factor that will influence political conduct. War, as Clausewitz explained, has a durable and unchanging nature, and the forces at play are embodied in the trinity of passion, chance and reason. While the fundamentals associated with the subjective nature of war are liable to change, at all times, the objective elements like danger, exertion, uncertainty and chance are inalienable from war. The political, military, social, technological, economic contexts inter alia, are the main determinants of the character of war, which, when defining the forms of combat, is referred to as warfare. War is, thus, a form of cultural behaviour and reflective of the character of the participant communities and whether or not a particular community feels threatened, depends on the particular political context it perceives. The Indian concept of limited war cannot, therefore, be divorced from its political context.

The fundamental problem of prosecuting a limited war is rooted in the clash between two definable paradigms of military security: One, in which the organising construct rests on preparing to fight and win the war (conventional paradigm), and another, in which an avoidance of war is the underlying goal of military preparations (nuclear paradigm). Paradoxically, both the nuclear and conventional paradigms of war coexist concurrently even though they do so uneasily.

This fundamental discord pervades all levels but more so at the political, strategic and operational levels. Planners are, therefore, confounded as most military plans tend to carry within them the seeds for mutual suicide. Limited application of force for limited purposes becomes the preferred norm, but it is a preference that requires cooperation from an adversary in times of political tension when distrust and suspicion hold sway and chances for miscalculation and misinterpretation of the adversary's action would be at their peak. The costs of failure and nuclear use are inestimable and the colossal scale and speed of destruction could even pose an existential challenge. The ideas of intra-war nuclear deterrence are actually forms of military fictions that will not withstand the scrutiny of the practical realm. Military leaders have no choice but to evolve operational concepts that supposedly minimise damage to oneself and could buttress deterrence by terrifying the adversary. A frightened, alarmed and panicky adversary then reacts to counter the threat and inevitably a chain of action–reaction cycle sets in that involves more arms acquisitions and leads to the creation of esoteric operational concepts like 'follow-on forces attack', 'anti-access area denial' and 'Cold Start'. India and Pakistan could learn from the

Cold War and avoid the unnecessary scaling up of arms procurement in the name of deterrence. But such avoidance can only come about through an improvement in political relationship that is based on the political leadership's acceptance. While force application may change the status quo temporarily, it will retain unresolved political issues that can seed future conflicts.

War is, therefore, increasingly being replaced by threats and brinkmanship. Since nuclear weapons directly impact the minds of decision makers, they produce strategic effects on the course of events in the form of 'deterrence action'. These 'virtual' engagements take place in the minds of adversaries. This is reflected as brinkmanship which emerges as a substitute to war-waging. Thus, the Clausewitzian notion that strategy is about 'engagements' for the purposes of war finds realisation only as long as deterrence holds. If deterrence fails and if force application follows, the Clausewitzian idea of force being directed at the centre of gravity, cannot be the preferred course of action due to the risk of uncontrolled escalation. Application of force will have to be circumscribed in terms of quantum, geographical spread and political objectives.

The post-nuclear India–Pakistan strategic interaction continuum fits Rupert Smith's theory of the altered paradigm regarding the use of force in the modern age, in contrast to what prevailed in the nineteenth, and most of the twentieth century,[5] when wars were fought by crushing the adversary's armed forces in an annihilating battle, thus forcing the collapse of the state. Winning the trial of strength was expected to subdue the will of the opponent.[6] This concept was basic to that age which Smith describes as the Industrial Age War wherein the paradigm followed a sequence of peace–crisis–war–resolution, with war consisting of military action as the primary

---

[5] Smith, Rupert. *The Utility of Force*, Penguin Books, London, p 181.

[6] Ibid, p 59.

deciding factor. With the advent of nuclear weapons, wars and the use of force was directed at the will of the adversary's populace and termed as 'war amongst the people'. During the Cold War, the armed forces were structured and could be amassed for employment within the old paradigm. In reality, however, no force was ever employed, especially at the strategic level and confrontation never became a conflict. The new paradigm, on the other hand, follows a different sequence consisting of confrontation–crisis–conflict–confrontation with confrontation being the enduring feature of the relationship and one that is characterised by unresolved political disputes. The armed forces are available at short notices and trained to fight wars, which cover the continuum from low-intensity to high-intensity wars, including those involving the use of nuclear weapons. Confrontation, which is the default mode, is sporadically visited by crises that could transmute into conflicts and involves the use of limited military force for limited political purposes. Such conflicts do not resolve political disputes in an enduring manner but instead revert to confrontation with perhaps an altered status quo. Smith's paradigm of confrontation–crisis–conflict–confrontation fits the India–Pakistan relationship in the way it has existed after the advent of nuclear weapons as conflicts have been fought with caution and remained geographically confined, primarily to Jammu and Kashmir.

There is also a clash between Indian and Pakistani beliefs regarding the utility of nuclear weapons in the ongoing India–Pakistan strategic process. Pakistan perceives its utility in deterring conventional and nuclear war. India considers its utility to deter only nuclear war. This clash of perceptions thus results in Pakistan attempting to leverage terrorism under the cover of nuclear weapons while India responds with the concept of limited war. This clash will not lead both nations into inactivity, rather, the probability of war diminishes and the probability for brinkmanship increases. Brinkmanship, however, tends to treat escalation as a strategy and bestows undue faith in deterrence.

Deterrence and Brinkmanship

Historically, conventional deterrence, though a preferred tool of military strategy, has been unpredictable, since strategy and plans can be constructed to circumvent the adversary's own strategy and plans. Nuclear weapons induce fear of survival; human behaviour in such circumstances is unreliable for rational thinking and erratic for inducing actions that one expects the adversary to take. It is also likely that decisions would have to be based on the perception of the adversary that is skewed in terms of reality, due to misinterpretation, misperception and information deficit or overload. If decisions were to be taken from a nuclear protected underground bunker, it would add an ambience that cannot but impact the leadership in unpredictable ways. This shaky psychological foundation of nuclear deterrence presents a danger that is best circumvented. Deterrence can also fail due to inadvertence and accidents. Brinkmanship based on a misplaced faith in deterrence is, therefore, an extremely risky undertaking that could be justified only by the vital interests at stake. It would, therefore, be foolish to indulge in brinkmanship without preparing for the consequences of its failure.

The operational challenge for nuclear forces is survivability, including its command and control mechanism. The size of the arsenal must be designed for survivability but the natural impulse to seek survivability in size must be resisted. Here, the lessons of the Cold War are pertinent and a residual capability to strike a limited number of significant targets is sufficient to deter, as calculations are normatively based on damage to one's own, rather than to the adversary, as deterrence relies on the answer to the question: 'What damage will I suffer?' Ultimately, all deterrence is self-deterrence.

NOTION OF VICTORY

'No one starts a war,' wrote Clausewitz, 'or rather, no one in his senses ought to do so, without first being clear in his mind what he intends to achieve by that war and how he intends to conduct it.' Thus, wars are fought to attain certain political objectives and decide specific issues. Technology has always influenced the character of war. But with increased destructiveness of military technology, the speed at which destruction is feasible, its wider availability to participants, it has become increasingly difficult to achieve a decisive military victory that fulfils the need to completely subdue the opponent's will through application of force without also suffering substantial damage to oneself. Of course in theory, it is feasible to completely destroy the adversary if an overwhelming number of weapons are used in a 'bolt from the blue attack'. Such a scenario is not pertinent to the Indian context as the size of the country and the impact on neighbouring countries would preclude such a possibility. With the addition of nuclear weapons in the India–Pakistan context, decisive military victory can well entail political risks that are not worth the end state. In any case, military victories do not themselves determine the outcome of wars; they only provide opportunities for the victors; even these opportunities could be limited by circumstances beyond their control. The outcome of wars is finally decided on the peace table where the fruits of military success must be encashed to achieve the political objectives for which the war was waged. As stated earlier, in the India–Pakistan context, at best, the outcomes may change the status quo but not necessarily resolve political issues in a manner that is likely to endure. The fundamental unresolved challenge of military planners in India and Pakistan remains the evolution of military plans that will provide a reasonable chance of being confined to the conventional realm and yet, achieve political objectives that justify the risks. Is India's search for conventional space a feasibility that can withstand military realities, and if there is space, what political objectives can it deliver?

POLITICAL OBJECTIVES

Political objectives are a product of interests which need protection and promotion and involve the creation of power to resist transgressions of interests. Military power backed by economic resources must be aligned to match perceived threats to interests. In the India–Pakistan context, neither country can apply military power to subjugate each other's will. A visceral sense of hatred is built into the DNA of significant sections of both nations and neither can afford being seen as having sacrificed any interest without substantial gain. Political solutions and compromises have not only been difficult to achieve but also even more difficult to sustain.

The conflict of interests could produce situations where use of force is threatened or applied. In the nuclear era, the use of force cannot accomplish the subjugation of the enemy's will; instead, force can at best be used to cause limited destruction and capture some territory that can be used for bargaining purposes. Prior to the nuclear era, Indian military power was essentially shaped to defend its own territory and capture a maximum of Pakistan's territory. This was its military objective and it was grounded in the idea that captured territory would provide bargaining power at the negotiating table. Despite the advent of nuclear weapons, India's earlier military objective has endured even as the political objectives have got altered. As the alteration has happened psychologically, it has not pervaded the political guidance provided to the military. Political guidance, as it exists, has not articulated with clarity the political objectives and this deficiency opens up the space for varied interpretations by individual services that have carved out service-centric plans for their growth.

POLITICO-MILITARY INTERACTION

The articulation of political objectives must guide the shaping of military power and the conduct of major operations. The evolution of political objectives that are to be achieved through threat or the use of military force must be arrived at through a process of institutionalised interaction between political and military leaders. The perennial danger lies in the military instrument being asked to accomplish tasks beyond its capability or else its potential when they are sub-optimally utilised. The problem is rooted in the nature of war that is inherently unpredictable as it is a multilateral interaction that is driven by forces that are not easily amenable to moderation by either party. These forces pervade both sides and possess far too many unpredictable variables that span the human emotions that violence unleashes, the interests at stake, the quality of leadership, material availability, weather and terrain, volatility of external support and sheer providence.

The Indian military leadership has no alternative but to plan for the entire spectrum of war to cater for failure of deterrence and ensure credibility; what is thus imperative is the need for clarity regarding the linkages between military force, political objectives and risks involved. Political objectives that are perceived to lie along the higher risk spectrum should be communicated to political decision makers through an institutionalised interaction. One should also be wary of the fact that political objectives with which belligerents start wars do not stay fixed during their course. Furthermore, the means adopted by India will be determined by those adopted by Pakistan and vice versa. There is, therefore, a large degree of uncertainty that cannot be wished away. Since there is also a substantial international community factor, the uncertainty is further magnified. The need to crystallise political objectives through an entrenched process of regular interaction between political and military leadership is recognisable and indisputable.

The contours of conventional space must be sketched by political restrictions within the ambit of which the armed forces should plan their operations. Such restrictions will call for a continuous review;

military planning will have to get accustomed to being kept on a tighter political leash. Understanding risks of operational plans is essential for political decision-making regarding war. The interaction must facilitate an understanding by the political leadership of the utility of the military instrument in various contingencies: provide an exposure to the risks involved; enhance understanding of capabilities that are required to be built; and evolve political outcomes that are considered feasible by the extant potential of the military, against the background of the adversary's capability. It would be obvious that only by institutionalising the process of interaction can there be a sustained dialogue that is able to incorporate developments in the political, strategic, economic and technological domains. A periodic strategic defence review is therefore a strategic compulsion that must be undertaken periodically.

The extant structures of national security decision-making of which the Cabinet Committee on Security (CCS) and the National Security Council (NSC) provide the platforms, are suitable mechanisms for institutionalising politico-military interaction. However, the absence of an integrated military perspective through the chief of defence staff (CDS) or any other such institutional head robs the interaction of a coherent military view that integrates the tri-service viewpoints. With the existing arrangement, the political decisions are based on individual service views with the chairman of the chiefs of staff committee (COSC) wearing two hats, normally finding it difficult to transcend the viewpoint of his own service. The absence of a CDS for a nuclear power like India is an unforgivable structural deficiency that can brook no further delay. Imagine the availability and feasibility of COSC to provide integrated military advice to politicians while directing his service in an ongoing conventional conflict. To expect that a chief will leave the conduct of operations to the vice chief is impractical as only a CDS can resolve the issue.

The politico-military interaction during times of crisis would understandably be deepened and intimate, but the important point is that the learning process regarding the military instrument cannot commence only at the start of the crisis. For the pace of events and mental stress may hinder, to a considerable extent, a better understanding of the issues thus impacting the quality of political decision-making.

The problem of institutionalised interaction has its origins in the nature of political decision-making as decisions are normally focused on the contingency at hand and viewed in terms of the immediate and short term. Political outcomes are malleable and shaped by circumstances, which are variable in a crisis and conflict. The military seeks clear-cut objectives and outcomes to be given to them by political leadership and such clarity is seldom feasible in a situation where the political leadership is trying to handle the domestic and international polity, as also one or multiple adversaries simultaneously. Political logic often trumps strategic logic and perhaps rightly so. But, if military leaders lack an understanding of the contingent nature of political behaviour, this will prove frustrating. While a natural tension is bound to exist, but its minimisation demands a deepened politico-military interaction, even if it does not always deliver an agreeable solution.

Conventional Space

Officially, India acknowledges that there are definite limitations to the conventional space available for the prosecution of limited war. The perceived space is a unilateral mental construct based on a judgment regarding the rationality of the adversary. But the theory of limited war is not easily amenable to operational solutions due to the fear of escalation and is therefore, understandably, hostage to the clash of conventional and nuclear paradigms.

The nuclear threshold will remain a nebulous mental construct, which is impossible to determine unless it is breached during a crisis. The nuclear factor must be taken into consideration while crystallising political objectives as force application will be circumscribed by politically imposed restrictions. This, in turn, would be determined by the interests at stake, the degree of risk aversion and the Indian reaction to Pakistan's military leadership playing the nuclear card.

The nature and scope of political objectives determine the expanse of conventional space required by the military. The moot question will be whether the perceived conventional space is sufficient to meet the military's need to ensure achievement of given political objectives which themselves could change during the course of the crisis/war. In theory, political objectives supposedly drive strategy and the resources required. In practice, once war commences, the uncertain course of war tends to privilege military priorities and political objectives become contingent in character.

Substantial political objectives require larger conventional space which, in turn, increase the risk of advertent or inadvertent escalation into the nuclear realm thus posing considerable difficulty in achieving significant political outcomes through war. Pakistan's nuclear threats, aimed not only at India, but also designed to impact the international community, particularly the US, could also act as another dynamic in determining the extent of conventional space that is hampered by the lack of political space. It may also likely be the case that risk perceptions of political leadership would be far more conservative than the military leadership.

During the Kargil Conflict, conventional space was restricted unilaterally by India's political decision makers by limiting the political objective of getting Pakistan to withdraw. While, there is no doubt that conventional space exists, there are reservations and uncertainties as to the scope of political objectives that can be achieved and this dilemma was illustrated during the India–Pakistan crisis of 2002 when the futility of the 'big fight' became obvious though the feasibility of 'limited war' endured and limited force for limited purposes became the accepted norm.

Military Strategy and Plans

The question of how any of the military options could be exercised to achieve the political objective of getting Pakistan to stop supporting cross-border terrorism and yet, remain within an envisaged conventional space requires examination. It is a vital question whose answer must offer convincing possibilities for the political leadership when the decision for force application is being formulated.

Surgical air strikes on known terrorist targets in Pakistan have for long been touted as a suitable response to another Mumbai-type attack. How such strikes will stop Pakistan from supporting terrorism, defies logic. For instance, Pakistan has not stopped supporting the Afghan Taliban, despite the sustained use of drones by the US to target Afghan Taliban leaders in Pakistan's western border and also killing innocent Pakistani nationals in the process. To expect Pakistan to act differently as a reaction to Indian military strikes would merely be wishful thinking.

However, the most often-used word in the Indian debate is that our reaction should be 'punitive'—to inflict pain through punishment, which in turn is inflicted by conventional explosive power. This, it is hoped, would at the least give Pakistan room to rethink, keeping in view the damage it will suffer. This argument ignores a vital piece of reality: that as long as Pakistan retains free will that is not deterred by an Indian threat, it could retaliate. An exchange of firepower would eventually be tantamount to a futile exchange of energy that reinforces historical animosities but resolves nothing politically. However, it serves the domestic audiences on both sides and satiates, however unsatisfactorily, their thirst for revenge and perhaps acts like a safety valve that helps to contain a long-running feud within the bounds of acceptability. In any type of India–Pakistan military conflict, the notion of victory will, therefore, be based on who has suffered more damage. But such an outcome would be disconnected from the original purpose of the venture—getting Pakistan to stop supporting terrorism. It could even be assumed that an Indian military strike would serve to unite all the jihadi forces in Pakistan and, in the long term, strengthen the very forces that are required to be weakened.

How much punishment would be required to be inflicted on Pakistan to stop it from supporting terrorism? Will substantial destruction of Pakistan's armed forces be enough? Such a scale of destruction, even if achieved within the bounds of conventional space, would also mean internally weakening an institution that could prevent the takeover of the state by jihadi elements—a paradox that makes Pakistan's armed forces simultaneously both the perpetrator and saviour. Of course, there is also the possibility, however remote, of jihadi forces taking over the armed forces. Any jihadi takeover is definitely against India's and world's interests and belongs in the gallery of nightmare scenarios regarding Pakistan.

The capture of territory is a favourite military objective and quite understandably the ultimate rationale for the Indian Army's operational plans only because possession of Pakistan's territory at the end of the war is expected to provide the bargaining tool in post-war negotiations. The logic behind this argument is based on the assumption that Pakistan will give up terrorism to get back its lost territory. The fallacy resides in the belief that Pakistan would accept that it was using terrorism as a tool of foreign policy and that even if it agrees to refrain from terrorism and thereby regains its territory, it will thereafter keep its promise. Alternatively, if India has captured substantial territory, Pakistan may prefer to launch an insurgency in that territory—an effort that suits its capability and style. The idea of capturing large parts of territory will run the risk of holding on to it against small groups of jihadis who will utilise the products of the lethality revolution—IEDs and suicide bombing as their primary weapons. It will also facilitate the expansion of its arena for furthering its strategy of a thousand cuts. Such empowerment of small groups has brought about a revolution in military affairs that has forced the exit of the most technologically advanced and militarily powerful nation in the world, such as the US, from Iraq and Afghanistan in recent times. The idea that technology is the sole arbiter between victory and defeat has been disproved and, instead, the will of the people has endured as the prime determinant.

Control of territory requires that eventually one has to have men on the ground with arms. The man with a gun who guards lives and material is indispensable to the control of territory. If the insurgent is able to hide amongst the population because he enjoys their support, he can, over an extended period, with very basic tools of technology, the AK-47, IEDs and suicide bombing, cause damage much beyond his size. The only way to counter such groups is to deprive them of the people's support; a fact that has been at the core of the Indian Army's counter-insurgency campaign in Jammu and Kashmir and the North East. But in the India–Pakistan context it would mean that people in captured areas across the India–Pakistan border will have to be won over (a practical impossibility) or those areas will have to be depopulated either during the conduct of operations or subsequently. A cursory look at the population density along the India–Pakistan border in Jammu and Kashmir and Punjab would provide a glimpse into the magnitude of the problem. Combat could easily be overtaken by the sheer size of the humanitarian problem as a result in terms of the population displacement. In the areas opposite Rajasthan and Gujarat though the density is lower, yet it cannot be ignored as a factor in the conduct of war.

The capture of large tracts of territory as an Indian military objective could be replaced by multiple shallow thrusts along the entire border. Such shallow thrusts largely resolve the problem of protecting territory as long as the captured area is either cleared of its population or is in size which can be easily defended against infiltration or conventional attack. In most areas, thrusts could be aimed at improving our defensive posture by utilising water obstacles and dominating terrain features. But capturing territory, for political bargaining to stop terror, still runs into the problem of Pakistan reneging on agreements or promises, especially since its practice of terrorism has deniability as its hallmark. A further aggravation is the fact that home-grown jihadis in the form of the Indian Mujahideen (IM) and other groups have manifested themselves and though they are supported and directed from Pakistan, it can claim that terrorism experienced is India's internal problem for which they are being unfairly blamed.

Accepting the notion that the military instrument has limited utility in resolving the cross-border terrorism issue will require discarding shibboleths that have barred the acknowledgement of the impact of the twin revolutions on the use of force in the India–Pakistan context. The role of the military must shift away from the notion of capture of territory as its primary objective to the ability to retaliate with firepower to counter similar activity from Pakistan. Although, defending India's territorial, maritime and air space would remain the primary role, the means of defending must modify from the existent 'capture of territory' focus to 'destruction' as a prime means of deterrence; not only because it offers a military solution to the issue, but also because it provides at the very least, an avenue to vent anger.

The transposition to destruction does not, by any means, imply giving up capability to capture territory but instead lowers its existing pervasiveness in an operational outlook to privilege destruction as the prime means. Capture of territory also ranks higher in the escalation ladder than exchange of firepower. Long-range firepower vectors in terms of air power, missiles and long-range artillery along with reconnaissance and surveillance capability would be required.

India, therefore, needs to develop capacities to hit back routinely with an intent to hurt. To adopt such an approach, it needs to restructure, so that it can punitively hit Pakistan without posturing. Necessary restructuring and modernisation of military assets, therefore, need to be carried out alongside ensuring a faster institutional response.

A similar rethinking, with suitable modifications to defend our border with China, would be a key requisite with the focus being on leveraging the power of defence in mountainous terrain, and also, especially, interdicting through firepower and other means the long lines of communications that China has to depend on, to project force across the Himalayas. The strike corps sanctioned in 2013 for the mountains, though publicised as being part of Indian military's riposte capability in case of China launching a major invasion, will require a re-evaluation. For, it is founded on a deeply-held belief regarding capture of territory that has been transplanted from the plains of the western border to the high-altitude mountains of the north, where defence has to be a superior operational technique. It must be understood, however, that limited offensive capability is an intrinsic component of defence in the mountains. In any case, the idea of India's military projecting power across the Himalayas onto the Tibetan Plateau ignores the creation of logistic vulnerabilities that could quite easily be exploited by the adversary. A rethink on the issue requires an examination of the basic assumptions on which the logic of a mountain strike corps is founded.

More importantly, the military must first focus on accumulating capabilities to achieve limited military objectives for limited political outcomes instead of attempting to get ready for the 'big fight'. There must be recognition that the presence of nuclear weapons imposes caution on political decision makers, despite the rhetoric often heard during crises.

Amongst the military, the Indian Army, in particular, has to make the major shift from its present thought process that privileges capture of territory over destruction. A shift to a firepower-based 'destruction' approach should force a major restructuring of the Indian Army's striking power that is now resident in three strike corps but whose relevance has been degraded with the conflation of the twin revolutions in military affairs.

Military strategy cannot escape the shackles that are a product of the clash between the nuclear and conventional paradigms. Risk of escalation must be catered for and the best must be sought from what is definitely an awkward situation for military planning. This, therefore, would entail planning for continued low-intensity conflicts like the proxy wars in Jammu and Kashmir, punitive strikes and geographically contained limited wars. Attempts to develop full spectrum war capability must be tempered with resource availability and assessed probability of threats along the higher spectrum of war. There is also a dire need to reconcile the divergence of views with respect to the impact of nuclear weapons on limited war between the services, especially the army and the air force. A joint doctrine could perhaps resolve the issue.[7]

Military strategy requires integrated thinking and coordinated application between the three services. Integrated evolution of strategy is hampered by the structure of India's higher defence leadership. The unsuitable arrangement that exists for inter-service cooperation in the form of the chiefs of staff committee (COSC) has been widely acknowledged to be dysfunctional for any major integrated decision-making. Several committees have proposed a change by creating a chief of defence staff, but such proposals have not found favour with the political leadership abetted by resistance from the Indian civil bureaucracy. The latest committee headed by Naresh Chandra has recommended a permanent chairman of COSC. According to

---

[7] A joint doctrine for the armed forces was released by the COSC in the last week of April 2017. The document has been widely criticised as falling short on several counts and certainly does not doctrinally reconcile divergent views within the services.

newspaper reports, the CCS considered the report on 2 May 2014 and did not approve the key proposal of the permanent chairman of COSC citing lack of consensus among political parties for the ten-year-old proposal for a chief of defence staff.[8] The issue while said to be under review by the NDA government, is most likely to have been quietly put on the back burner. Coordinated application of military power is also handicapped by the absence of integrated commands, which is also a need that has been acknowledged but reform has remained hostage to the absence of a CDS and the vagaries of inter-service incoherence.

Currently, the role of the Indian armed forces in several ways remains misaligned to the country's political objectives and its propensity to pose an increased threat to Pakistan is counterproductive. The shift to a proactive strategy of being able to launch its offensive ground forces in a short time frame, in order to overcome the reaction time of Pakistan, is operationally sound, but fails in the realm of strategy. It creates the effect of an increased Indian threat which plays into the hands of the Pakistani Army that uses such a threat to strengthen itself within Pakistan's political firmament.

Strengthening the Pakistani Army with an Indian military threat encumbered with the complications of escalation merely delays and reduces the odds of strengthening the moderate constituency in Pakistan. India's interest in strengthening the moderate constituency in Pakistan is based on the belief that Pakistan's salvation from the grip of terrorism, that the state itself has spawned, has to be confronted by forces within Pakistan, whose spearhead would have to be the politically moderate constituency. India's strategy of dealing with Pakistan's terrorism will have to be aimed at strengthening such a moderate constituency and one of the major prongs of the effort has to be reducing the visibility and form of the Indian

[8] *The Indian Express*, 3 May 2014, www.indianexpress.com/article/india/india-otheres/govt-rejects-chandra-panel-proposal-for-a-four-star-general/

military threat. To do so, military logic would have to reorient itself to serve political purposes and not be based on the logic of seeing the demise of Pakistan as a state but rather see its transformation as a friendly and stable neighbour. No doubt, political leadership on both sides will have to work assiduously to disabuse their populace that India–Pakistan differences can be resolved through the use of force, in any form, especially the conventional or sub-conventional. The main factor for embracing such a notion is embedded in nuclear weapons that both nations possess. It now requires a leap of faith whose time has come.

These realities have not yet entered the debate in the Indian strategic community, for very often the military part of the debate is safely straitjacketed at the operational level and seldom ventures to connect the larger issue of strategy and its linkages with political objectives. The military leadership must be able to connect military success to the goals of policy but is often unable to link the two. The absence of this linkage in military debates has its roots in India's professional military education system of higher leadership.

Strategic Communications

Absorbed information influences beliefs and shapes behaviour and the potential for information to influence mass behaviour has always been an enduring factor in the domestic and international political spheres. The competence to convey information and achieve strategic effect by influencing masses of population or the decision-making elite has been described in recent times as strategic communication. It involves the ability to communicate narratives to multiple audiences through media that capture the attention of the target audience. Whether in peace or war, confrontation, crisis or conflict, the nature of political interaction is shadowed by the influence of the information domain comprising in contemporary times a 24/7 media, including the social media and the mobile telephone that offer connectivity to an extensive world of information.

The main challenge of strategic communication will always be the countering of opposing narratives from multiple sources and the inability to influence the interpretation of the information by the recipient. In a democratic dispensation like India, a free and independent media could create a cacophony that could easily drown the voice of the government unless it is organised to communicate effectively. The voice of the government has to emanate from the respective ministries which deal with the subject and therefore, each ministry must have a structure for public information and since there are many cross-cutting issues, there has to be a platform for coordination between ministries. Presently, the structures which exist are grossly inadequate and very often reflected in information incoherence within the government. The problem is compounded by India's federal structure wherein states and Centre could be working at cross-purposes because of differences in perspectives and interests. This is especially problematic if the issue also involves an external actor like a neighbouring country. A system for coordination based on consultation and dialogue could minimise the problem even though the federal structure of India can be expected to remain resistant.

In August 2002, based on the Kargil Committee Report and the group of ministers recommendations, a National Information Board (NIB) for national level policy formulation on information warfare and information security was established. It was also responsible for the creation and monitoring of the required institutions and structures for implementation of the policies. The NIB is chaired by the national security advisor and serviced by the National Security Council Secretariat (NSCS). Its members include cabinet secretary, chiefs of the armed forces, secretaries of Foreign, Home, Defence, Finance, Information Technology, Telecommunications, Information and Broadcasting, and Space Ministries, heads of intelligence agencies, scientific advisor to the defence minister and the deputy NSA as member secretary. Nothing much is known of the functioning of the NIB except that it held several meetings during the tenure of the second NSA, JN Dixit, and that a National Security Information Policy was finalised for submission to the CCS.[9] However, it seems that NIB has been practically comatose for long, since none of its exertions seem to be known. Presumably a vacuum in this vital area of national security exists. It is apparent that in the absence of an effective NIB, incoherence in government communications is not uncommon, with the larger perennial danger of ministers and heads of organisation making statements that could vitiate a crisis situation and/or create avoidable tensions with other nations. Specifically, statements by politicians and officials on India's nuclear and military capability must conform to policies that the NIB must formulate and impose.

One of the main issues is the creation of institutionalised structures manned by specialists in the field of mass communication. Every ministry must treat strategic communication as an inherent requirement in the process of delivery of public goods. Every roll out of a government's policy or action must have a communication plan that packages its messages according to the characteristics of the targeted audience, which could be multiple in most cases, with transmission carried out through diverse mediums.

No war or conflict can be conducted effectively without a strategic communication plan that is built into the evolution of the plan for the conduct of the war or conflict. This requires structures that are already in place and have developed the capability to scale up, to meet the demands that are placed on them during contingencies like war or conflict. The communication links for coordination must provide the platform that harmonises differences and reduces the

---

[9] Chandra, Satish. 'National Security System and Reform' in *India's National Security Annual Review*, 2005, Knowledge World, New Delhi, p 213.

incoherence of the government. In an interconnected world, such links are not difficult to establish and leverage.

A culture sensitive to the importance of strategic communications must pervade the government machinery and be part of the training curriculum of government officers. Political recognition of the importance of institutionalising strategic communication is an obvious primary requirement. But, surely, strategic communication is an area that implores for greater attention than it has been given so far. An effective role in the international arena requires an efficient strategic communication infrastructure that is able to facilitate the achievement of dominance in the information domain. Ultimately, unless the battle of opposing narratives amongst multiple audiences is won, physical victory cannot be successfully translated into a lasting peace.

The Indian nuclear doctrine is a good example of strategic communication. It conveys with clarity India's intent, capability sought, conditions for use and the type of command and control to be exercised.

Professional Military Education

India's professional military education system for higher leadership was mostly established post-Independence and retained the flavour of its British legacy. Institutions like the National Defence Academy, Indian Military Academy, Naval Academy and Air Force Academy provide entry-level training at the joint individual service and joint service levels. Mid-career joint service training is provided for the major and lieutenant colonel level at the DSSC in Wellington, which concentrates on tactical and operational level staff work. The Higher Command and equivalent courses at the colonel level expose the officer to the operational level command functions and are conducted at the Army War College, Mhow, College of Higher Defence Management, Secunderabad, College of Air Warfare,

Secunderabad and Naval War College, Goa. The National Defence College, New Delhi, conducts training on national security issues, with an institutionalised exposure to the military as an instrument of politics at the senior (brigadier) level with a healthy mix of civil services and officers from foreign countries. The main problem is that this exposure comes too late in careers of the defence personnel. Also, it makes the shift from the operational level to the strategic—an intellectual exercise that is not easy due to the calcification of ideas accumulated in nearly thirty years of service. The exposure to the 'nature of the whole' and how the military as a political instrument connects to political objectives is imparted only at the highest level in the training ladder instead of forming the backdrop at all levels. In the present system, due to late exposure, the outlook remains narrow with the focus on tactical and operational levels that is focused to expand the skills for applying the military instrument and seldom goes on to connect its impact to the arena of political purpose. Ironically, the military is culturally disdainful of politics and proud of the fact that being an apolitical force has been deeply ingrained from inception. But an understanding of politics and its nuances is necessary for the military to perform its tasks, especially for the higher leadership as all military force application or threat is primarily for political effect. In counter-insurgency operations, such understanding is even required by junior leaders whose actions directly impact the centre of gravity of the military effort—the people. The primary challenge to generalship is mastering the art of translating the conversion of military effect into the desired political impact. It requires a currency conversion that has no fixed exchange rate in scenarios of uncertainty and chance. The exchange is dynamic and is captive to the politico-military context. At the best of times, it is an unenviable task and without an understanding of the political conditions it could become self-defeating. It is this lack of understanding that has challenged the military leadership from debating the question as to what mattered the most after the India–Pakistan nuclearisation—what is the role of the military instrument in getting Pakistan to stop its support of terrorism?

The intuitive answer to this question is building greater military power that can be used to threaten or be applied to coerce Pakistan to end its use of terrorism against India. In fact, this has been the trajectory of India's military expansion since the 1980s and continues unhindered except when slowed down by the emergence of China, defence budget constraints and problems with our defence acquisition system. But the military needs to review some of the shibboleths that it continues to harbour despite changes in the objective conditions consequent to nuclearisation and the need to align military objectives with political outcomes.

Military Rebalancing

The strengthening of Chinese military power in the last two decades, especially its creation of military infrastructure in Tibet, has forced India to strengthen its military capability on its northern borders. The need for developing such a capability gained recognition during the last decade and has seen the creation of additional forces for the army and air force. Two divisions have already been raised in the North East and a mountain strike corps was approved in late 2013. The air force has also deployed some fighter aircrafts in the existing bases in the North East and plans to operationalise several forward operating bases along the Indo-China border. The negligence of our northern border with China was also a resultant of our focus on Pakistan as resources were inadequate to manage both. Presently, resources availability has reduced because of the shrinking of our economic growth; there is a resource crunch and yet, defence planning continues to aspire to create additional military power through new raisings. Such a trajectory of expenditure is likely to weaken the entire defence system since the paucity of resources will have to be shared.

India's land and air power, about 60 per cent, is oriented towards Pakistan and, worse, much of it is unusable and retain their logic for existence on deterrence of Pakistan which is in no position to attack India conventionally. This is so, not only because its adopted means is terrorism, but also because for more than five years nearly 60,000 troops have been deployed to fight home-grown terrorism towards its western borders. The size of our land power against Pakistan also draws nourishment from operational concepts that belong to the pre-nuclear era of capturing large tracts of Pakistani territory as a bargaining tool. The rational alternative is to rebalance towards the Sino-Indian border from the west. This requires the Indian military leadership to acknowledge the need to rebalance and develop a strategy that reconciles the means against the ends.

Alert Levels

It could be reasonably assumed that any future crisis would have its share of nuclear rhetoric that will impact in varying degrees both the domestic audiences and the political leadership. The media could be expected to drive jingoism and inflame passions, intensify hate and call for action that will assuage their emotional upsurge.

The real danger lies elsewhere. It could well be in the decision of the political leadership regarding a change in the alert levels of nuclear arsenals. Such changes will be driven by another emotion—fear. Contemporary crisis situations are usually encompassed in a stream of intelligence and information from a variety of intelligence agencies including shared foreign intelligence, diplomatic channels and open sources that comprise a plethora of media types. It would be happening in an atmosphere of suspicion, anxiety and uncertainty. The situation being pregnant with misinformation, misunderstanding and misinterpretation is a natural breeding ground for fear. If fear is a result of suspicion about the adversary having made or making plans to make nuclear moves, the pressure for similar reaction will be difficult to resist. The barrier of movement for deployment is a red line that should actually be a threshold that must preferably not be crossed. Deployment would mean weapons systems being locked into an action–reaction framework. This equally applies to nuclear forces deployment during ongoing conventional level hostilities where the danger is even greater as the ability for restraint could be surpassed by the pace of events on the battlefield. The events of the battlefield could dictate a change in political objectives and the tail could wag the dog. Such situations are not uncommon during hostilities but acquire lethality when nuclear weapons are involved. Avoiding deployment provides greater assurance against miscalculation. But it could be argued that there is room for pull back even if nuclear weapons are readied for use awaiting further orders. Such was the situation in several crises during the Cold War and most notoriously, the Cuban Missile Crisis. While the argument is right about the outcome, it does not take into account how close the situation was to a nuclear exchange. The issue is one of avoiding the situation rather than using the situation to manipulate risk that one cannot control, due to the fact that it depends on an action–reaction sequence in an ambience of mistrust, misperception, fear and uncertainty.

Mathematicians, economists and academicians have evolved a game theory to demonstrate how risks can be manipulated. These have become the staple of strategic education the world over. The problem with such postulations is that they are divorced from the realities of crisis situations. Game theories presume that cool-headed strategic thinkers represent the political leadership and that they have a fair amount of information about what is happening on the other side of the hill. These presumptions are grossly misplaced. In the difficult moments of a nuclear crisis, the leadership would, depending on the level of alertness, be underground in a nuclear-protected facility. The limited confines of space and the ambience of the underground facility itself will be stifling enough for the political leadership to be anything other than cool. Instead, it will be tense and uncertain leaders who will be attempting to make decisions that could have as their price of failure the very existence of the state. Even more, the lack of information or information overload could make it a guessing game that prevents the maturing of shrewd calculations needed to win the game. Games are good in theory but merely mislead and impractical for application in the real world. At best, they can be planning tools. In any case, politicians are not naturally imbued with strategic insight except, of course, they mostly come with sufficient common sense to not get carried away by advice that they know intuitively to be impractical. Probably, this is the reason why military advice on the use of nuclear weapons was ignored in all crises during the Cold War. Douglas MacArthur, of course, was even sacked for his unsolicited advice on the use of nuclear weapons in Korea. The avoidance of change in alert levels remains the main bulwark in the preservation of stability. Nuclear signalling, through change in alert levels, is a dangerous road that is best avoided.

In the India–Pakistan context, the continuance of effective and official back channel communications that is authorised at the highest political levels and insulated from the existing political tensions could probably serve to provide a better interpretation of nuclear signalling.

The issue of de-mated posturing will rise to the surface once India operationally deploys its nuclear submarines. The Indian nuclear posture would no longer be wholly de-mated. Pakistan can then be expected to come up with a weapons system that will also not be de-mated. China has nuclear submarines and is in the process of developing its long range SLBM. Deciding the pattern of patrolling for India's SSBNs will be a political decision that will demand a greater understanding of the nuances of deterrence from the Indian political leadership.

Major Conflict Avoidance

In a situation where both India and Pakistan are armed with nuclear weapons, a major conflict avoidance must be the political objective as was acknowledged in the Lahore Agreement. Ironically, at certain thresholds in political affairs, force could become part of the political discourse. The sword of the statesman will persist, even with the knowledge that operational techniques and military virtuosity offer no pragmatic solutions to force application in the nuclear age. Practically, punitive strikes and geographically confined skirmishes emerge as the political preferences, albeit, with a questionable ability to produce a substantial and enduring strategic and political effect of value which can then provide a basis for the resolution of India–Pakistan issues. Thus, the effect of military force application is confined to temporarily influencing the status quo but not to changing it significantly. Any major changes in status quo will involve a disadvantaged side and could be expected to revert and keep the pot boiling.

As both conventional and nuclear deterrence are known to be unreliable, India and Pakistan must, therefore, pursue both nuclear risk reduction measures and other confidence-building measures,[10] some of which could include:[11]

- Greater transparency on nuclear doctrine.
- Continuance of de-alert status even after the introduction of nuclear submarines.
- Abolition of short-range nuclear missiles.
- Avoidance and intimation of change in alert status of nuclear weapons.

---

[10] Currently, the only India–Pakistan agreement on nuclear risk reduction measures in place is the mutual notification of missile flight tests, signed in February 1999 and has been renewed on a five-yearly basis.

[11] Extensive research has been undertaken in the domain of confidence building measures (CBMs) and nuclear risk reduction measures (NRRMs). See www.stimson.org (It provides a good insight).

- Establishing risk reduction nodes to support implementation of measures agreed upon.

Dealing with Terrorism under the Nuclear Shadow

The threat or use of politically motivated force is inherent to the human social condition. India–Pakistan history holds no optimism that a positive kind of peace can be created to the degree that precludes the appearance of objective threats to security. Prior to nuclear era, war was a probable option, but under the shadow of nuclear weapons, any war, even a limited conventional war, has acquired high risks without offering any reasonable prospects of creating a peace that is sustainable. As a resultant effect of Pakistan's use of terrorism as a foreign policy tool, the use of retaliatory force may become unavoidable, the manifestations of which could be brinkmanship, punitive strikes, geographically confined conflicts and localised skirmishes. This, in turn, severely circumscribes the achievement of any substantial political objectives by the use of force and also poses major difficulties in the crystallisation of political outcomes desired through use of force.

Yet, both sides, frameworked within the security–insecurity dilemma and the stability–instability paradox under the nuclear shadow continue to explore for space albeit in different realms. While Pakistan seeks space through the irregular means of terrorism, India does so through limited conventional war. Paradoxically, the boundaries of the space being explored by both countries can only be known after the consequential conflict has crossed the nuclear threshold. Ultimately, it is an issue of the quantum of risks one is willing to take for the stakes at issue, even when one is not in control. Such a prospect has induced caution on both sides; a caution that has also inadvertently maintained a tenuous peace and remained a prime driver towards a solution through dialogue. The utility of force is restricted to protect the status quo—limited force for limited purposes, continues to be the practical though unsatisfactory option.

Post the introduction of nuclear weapons in the India–Pakistan strategic context, the force application paradigm has got transformed. The extant India–Pakistan paradigm is: 'Perennial Confrontation–Crisis–Limited Force Application–Perennial Confrontation.' The challenge, therefore, is to construct a lasting peace within the contours of the status quo and the notion of making the borders irrelevant is an idea that seems to carry some chance of success.[12] But this requires the embrace of a vision that promotes free trade and movement of people across the India–Pakistan borders which Pakistan has continued to resist. Considering the pervasive impact of geopolitics coupled with negative emotions which have historically imbued India–Pakistan relations, the challenge is indeed daunting. Most issues of contention remain insoluble and with sporadic spikes that are hopefully contained by the nuclear shadow, an uneasy peace is likely to be the default condition.

Both militaries have been in the pursuit of space below the threshold of major conventional war. Such a pursuit for space, though differentiated in segments, has continued to occupy the intellectual discourse on both sides and has not been able to evolve an acceptable pathway offering a reasonable chance of providing a force application solution to the political problems at hand.

The fragile peace which exists has essentially been contained through intermittent dialogues at the political level, a mutual agreement like the ceasefire agreement and the weekly director general military operations (DGMO) telephonic interactions. Overall, despite very little progress on pending political problems, force application has been largely confined within manageable levels. Considering the

---

[12] Pakistan had agreed not to redraw borders during secret negotiations in 2007. Statement by Satinder K Lambah, former Prime Minister Manmohan Singh's Special Envoy. Harinder Baweja, *Hindustan Times*, 16 October 2015.

level of the exchange of force over the four decades since the 1971 War, the international public narrative that the India–Pakistan border is one of the most dangerous trouble spots in the world is an overstatement. Although empirical evidence on force exchange indicates a frequent breakdown of the ceasefire agreement, there is also the fact that spikes of violence are always contained in time, space and intensity.

To end cross-border terrorism would remain a high priority for India and in the long term, this objective could be best realised, not by going to war, but rather, by engaging intelligently with Pakistan, by strengthening their democratic and moderate constituencies and taking effective internal steps to deal with terrorism and infiltration. The current idea of Indian force application rather than tackling terrorism from Pakistan serves to strengthen the forces of terrorism. It is only Pakistan who can endeavour to eradicate terrorism from within its polity.

However, the political leadership cannot be totally deprived of the instrument of force, should negotiations and pressure not succeed and for which restricted conventional space is available, that can, at best, temporarily alter the status quo. The political leadership must have at its disposal a utilisable military instrument, even if it is inadequate to resolve contested political issues. Thus, military preparations must prioritise feasible forms of force application. Seeking options for force application in a nuclear environment is the fundamental challenge for military planners. Quite obviously, punitive strikes without posturing is a capacity which needs constant strengthening. Also, privileging the creation of destructive capabilities over capture of territory is a long overdue doctrinal shift. More importantly, the military arsenal must be shaped and aligned to achieve envisaged political outcomes.

The greater challenge of setting goals for which force will be applied in retaliation to a major terrorist strike has to be confronted by India's political leadership. The goals set must be realisable by military force application and should be able to endure through the force application process. But since such feasibility is uncertain in a bilateral action–reaction sequence, the judgment of risks is inherently difficult, if not well-nigh impossible.

All the force application options possess the seeds of escalation possibilities. The dilemma, inextricably embedded into the decision-making process, will hinge on the clash between two issues. The first will be the call for action, revenge and restoration of honour from domestic political elements that could be expected to be magnified by the media catalysing the forces of jingoism. The second will be the acceptance within the political leadership that feasible force application quantum cannot resolve the problem of terrorism emanating from Pakistan. The nuclear revolution will stare at the lethality revolution and present a *dharamsankat* (moral dilemma).

India's Existential Threats

Potentially, India could face two existential threats. The perennial long-term threat, inherent in its post-Partition structure and derived from its historic legacy, is the communal fault line which can be threatened from communal forces of Pakistan, India and some other parts of the world. Pakistan's persistent attempts to widen the fault line have not met with success. Paradoxically, the worst-case scenario for the threat is the fact that through its practice of nurturing terrorist groups to attack India, Pakistan has become the catalyst of its own possible implosion. As stated earlier, if it collapses, a large-scale demographic flow into India would be impossible to stem and this, in course of time, could ignite India's communal fault line.

The other threat is a nuclear war between India and Pakistan, which, however remote a possibility, would, if it occurs, pose an existential threat to all countries in South Asia. Some studies indicate that an India–Pakistan nuclear exchange of about a hundred

Hiroshima type weapons would endanger the globe through climate change.[13]

In order to extract revenge, the notion that India should develop sub-conventional capabilities by leveraging the internal fissures of Pakistan in places like Balochistan, Pakistan Occupied Kashmir (POK), northern areas and North West Frontier Province is a questionable one. A 'quid pro quo' policy at the sub-conventional level is not only desirable for Pakistan but is tantamount to playing into its hands. Given that Pakistan is ahead in its capability to launch terrorist attacks successfully and India's sheer size, porous borders and liberal political ambience present Pakistan with easier options, the lethality revolution works in Pakistan's favour. The sub-conventional capability that India must develop is in the domain of information warfare, wherein it concentrates its efforts to gaining the people's trust by winning the battle of ideas. The cyber world offers the opportunity and coupled with India's intellectual and software designing potential, is strategically and politically the choice that suits us best. It would indeed be a long-term effort and offer no benefits for narrow domestic political parochialism that often privileges short-term benefits over long-term national interests. However, as that leaves India with narrowed choices in force application, maintaining the fragile peace that prevails in the region becomes the practical choice for contemporary times. Although, it is an unsatisfactory one, it is yet a wise one, given the circumstances. India and Pakistan must not confuse military and strategic success. The former is about defeating the forces of the enemy; the latter is about using the defeat to advance the goals of policy. In any case, the view of using sub-conventional capabilities to destabilise Pakistan is based on the erroneous presumption that a destabilised Pakistan should work

[13]Climate model studies based on smoke produced in an India–Pakistan nuclear exchange indicate a possibility of a global nuclear winter that could impact global crop production for decades. http://climate.envsci.rutgers.edu/nuclear/

in India's favour. In the long term, India could at best, expect the situation to change by supporting the moderate elements in Pakistan to overcome the vice-like hold of the forces of fundamentalism that have pervaded its polity. It might seem an impracticable goal but giving up such a goal is to condemn the India–Pakistan situation to one of perpetual strife. India's statecraft must be guided by its vision of creating a South Asian political and economic union which wages war on poverty, illiteracy and disease. Both the nuclear shadow and the lethality revolution provide little room for manoeuvre. The strategic choices India exercises must support its long-term interest of a stable Pakistan within a South Asian regional grouping.

No War Pact

The idea of a no war pact (NWP) has surfaced intermittently since the 1950s and on 22 December 1949, Prime Minister Jawaharlal Nehru handed over a draft NWP, to Pakistan's High Commissioner. Prime Minister Liaquat Ali Khan of Pakistan accepted the draft but stipulated an undertaking to 'resort to arbitration on all points of difference', but Nehru insisted that Kashmir was a political question that was not justiciable.

Pakistan then continued to insist that signing the NWP would mean accepting the ceasefire line. The agreements following the 1965 and 1971 wars included clauses on non-use of force, peaceful resolutions of disputes and after 1971, the inviolability of the LC. A NWP was rejected by Pakistan during the discussions prior to signing the Tashkent Agreement and the Simla Agreement in 1972, citing the issue of freezing the status quo in Jammu and Kashmir.

In 1981, along with an announcement regarding acceptance of a US military package and in the spirit of the Simla Agreement, Pakistan proposed 'mutual guarantees of non-aggression and the non-use of force'. India stipulated two conditions—bilateralism and eschewing bases or alliances with a foreign power. On 28 January

1982, Pakistan's foreign minister clarified, 'The proposed NWP applies to Kashmir and war is ruled out. Only peaceful means would be applied to the solution of the problem.'[14] Discussions on the NWP continued till 1987 but despite nearing resolution, fizzled out. From 1989 onwards, Pakistan has changed tack and employed other than-peaceful methods by orchestrating insurgency in Jammu and Kashmir.

The Siachen dispute had also approached a settlement during the five rounds of talks between January 1986 and June 1989. A breakthrough had appeared feasible in June 1989 but stalled due to opposition from various constituencies of both sides. Efforts at resolving the Siachen dispute have continued intermittently alongside other issues and the thirteenth round of defence secretary level talks was held in June 2012 without any positive outcome. There have been no signs of any flexibility on either side regarding the issue of demarcation of the existing troops' position prior to a withdrawal from Siachen. In April 2012, following an avalanche that buried 140 people, mostly Pakistani troops in the Giari Sector, the Pakistan Army chief had called for a demilitarisation of Siachen following a peaceful resolution of the issue. A resolution is unlikely to take place in isolation of other pending issues like terrorism, Kashmir, Sir Creek, Wullar Barrage, Tulbul Project etc.

Essentially, Pakistan has to commit to the non-use of force including its most practised form, terrorism. Both sides also have robust political constituencies that cannot brook any agreement without solving pending issues. Politically, it would be difficult to insulate the talks from continued infiltration across the LC, incidents of violence on the LC and terrorist violence in India. The NWP must mark Pakistan's realisation that India–Pakistan issues cannot be resolved by using terrorism as a tool of foreign policy, as the Indian

[14]Noorani, AG. 'Towards a No-war Pact between India and Pakistan'. South Asia Citizens Web, 21 December 2008, www.sacw.net/article439.html

state has successfully countered all such attempts. Neither the cover of nuclear weapons nor its sub-conventional efforts in the form of terrorism have produced any resolution of disputes in Pakistan's favour. On the contrary, Pakistan is on a dangerous trajectory of deepened militarisation to counter major threat from several domestic terrorist groups.

The present state of affairs was given a fillip by Zia-ul-Haq in the 1980s in particular when 'Islamisation, jihad and religious nationalism' became the staple impulses promoted by the state. By August 2012, and actually post-Lal Masjid operation in 2008, Pakistan's terrorism chickens started coming home to roost and the army chief was seen calling for the nation's support to fight terrorism and extremism and flagged the possibility of the nation being plunged into a civil war. By early 2014, following a series of major attacks on its armed forces and civilians, Pakistan had begun talks with the Tehrik-e-Taliban Pakistan (TTP). During this entire period, Pakistan continued to support selected terrorist groups that are focused on India and are against the West in Afghanistan. Running with the hare and hunting with the hounds has now brought the state to a crossroads where the fight against the behemoths of its own creation now has to be waged. Such a situation should force a rethink of its threat to India and therefore, the political conditions for a NWP, it was hoped, could crystallise post India's elections in May 2014. But all attempts at peace fell prey to continued cross-border terrorism. Without the moderate constituency in Pakistan succeeding against extremists, there is no hope of achieving stable and peaceful relations with Pakistan. India must support any move that strengthens this moderate constituency in Pakistan because the essential internal strife is between moderates and religious extremists. An NWP should strengthen the moderates in their fight even though efforts at NWP will be subject to sabotage by the extremists.

On the Indian side, the argument that we should not transform our military capability must be countered through acceptance of the notion of military capability becoming marginalised due to the nuclear equation, which calls for a restructuring aimed at transforming itself into a usable military instrument which could then make coordinated application possible through measures like establishing integrated commands. Both nations will have to free themselves from notions regarding the utility of force that has, so far, been taken for granted but in actuality has long outlived its potential and utility in the manner previously envisaged. Ironically in the India–Pakistan situation, force application does not weaken the adversary's political will but instead has the opposite effect of strengthening it. Pakistan has to part with terrorism and India has to relinquish the notion that military superiority is the antidote to terrorism. Coincidentally, the issue, though different for either side, involves the military elements of the divide. There is a convergence of interests that should provide the foundation to work towards an NWP even as the road to NWP will be cratered with the baggage of the past as much as of the vicissitudes of the present. Undoubtedly, the process will have to transit through several levels of agreement and restoring the sanctity of the 1993 Ceasefire Agreement should be the first step.

International Dimension of Conflicts

International support and cooperation form an essential constituent for the successful conduct and conclusion of crises and conflicts. This is because the political context of any war in contemporary times is inevitably imbued with an international dimension and it is challenging in the extreme for a nation to exercise its right to defend itself with unconstrained freedom of action. There are also limits to national sovereignty in waging war, which gives birth to a new outlook on the character of future conflicts. With the advent of nuclear weapons in the India–Pakistan political equation, international concern about India–Pakistan conflicts has got magnified and resulted in affecting the strategic autonomy of both states. Though India has steadfastly held the view that India–Pakistan political disputes must be resolved bilaterally, Pakistan has continued to prefer a third-party intervention. Even though most efforts at conflict resolution have remained bilateral, conflict and crisis situations have involved intervention by major powers. In the Kargil Conflict, the lack of international support for Pakistan, even from its normally steadfast ally, China, shaped the contours of the strategic realities against Pakistan. Pakistan courted the US, which helped in the resolution of the crisis. In the India–Pakistan crisis of 2002, India lacked international support in its threat to wage war, but the threat forced the US to pressurise Pakistan to make promises in order to prevent conflict. India's restraint after the Mumbai attacks focused the attention of the international community on the issue of Pakistan being the epicentre of terror. It is an image that has been reinforced by their experience and knowledge of several attacks that have either been carried out by terrorists with Pakistani connections or attacks aborted by intelligence agencies in many countries in the West. At a strategic level, it is its image and label of being a terrorist state that continues to be Pakistan's greatest loss in terms of its standing in the international community.

Conclusion

Pakistan's strategy of 'bleeding India through a thousand cuts' is juxtaposed against India's strategy of 'limited war'. The strategies are woven with the strands from belief systems founded on utility of force that are no longer useful to resolve the contentious political issues at hand. Both stratagems booby-trap their owners and the noose is tightened further by the conflation of the nuclear and the lethality revolutions. Escaping the trap requires the embrace of a new paradigm regarding utility of force and jettisoning the idea that superior and out-of-the-box strategies and operational methodologies can deliver political outcomes that will resolve India–Pakistan disputes. The utility of force does not evaporate but is instead marginalised to assuage domestic audiences, thirsting for revenge and at best temporarily changes the status quo without a lasting impact on resolving disputes at hand. Most importantly, with the degree and intensity of force application being circumscribed through the escalation factor, the impact on political will towards force application for revenge is strengthened and not weakened, which is the opposite effect that force is supposed to achieve. This is mainly derived from the cultural level where visceral hatred between both societies is the greater norm. On another level, it acts as an escape valve to contain popular anger and the desire for vengeance. Admitting, recognising and embracing this reality provide some space for escape from what are essentially strategies that are entrapped within the body of ossified beliefs. Politics, sans force, should be the preferred approach. This will require the political leadership on both sides to modify their attitudes towards existing ideas that relate to utility of force. The ability to change is also challenged by the domestic political benefits conferred by being perceived as taking a tough approach to a recalcitrant adversary. This too is no less a trap that can only be overcome through statesmanship on either side. While this is a possibility that may look remote, it does afford the greatest hope for peace under the shadow of the twin revolutions.

During the Cold War, deterrence was said to have worked because demands on it were limited. So, while there is limited scope for application of force under an India–Pakistan nuclear shadow, in the India–Pakistan context, it might well be the case that deterrence might fail if 'asked to do too much'. Limited war in the India–Pakistan context may not be born as yet and even if it does, whether it will survive its birth, appears doubtful. War is about peace, not about itself. Pakistan's adoption of terrorism as a foreign policy tool has not only been disastrous, but has now also become a threat unto itself.

The India–Pakistan pursuit of space for force application under the nuclear shadow is, in all probability, a road to nowhere. Thus far, it has mostly been a futile journey that has been textured by political expediency and military fiction, undertaken to temporarily satisfy the lust for vengeance. To continue along this path, is merely extending the 'stopover' in the strategy trap. An 'escape' calls for cooperation to be obtained by rising above historical baggage and the pervasive negativity of geopolitics, which only mutual political sagacity can deliver.

APPENDIX

Appendix A

India's Draft Nuclear Doctrine

- Preamble
    - The use of nuclear weapons in particular as well as other weapons of mass destruction constitutes the gravest threat to humanity and to peace and stability in the international system. Unlike the other two categories of weapons of mass destruction, biological and chemical weapons, which have been outlawed by international treaties, nuclear weapons remain instruments for national and collective security, the possession of which on a selective basis has been sought to be legitimised through permanent extension of the Nuclear Non-Proliferation Treaty (NPT) in May 1995. Nuclear weapon states have asserted that they would continue to rely on nuclear weapons with some of them adopting policies to use them even in a non-nuclear context. These developments amount to the virtual abandonment of nuclear disarmament. This is a serious setback to the struggle of the international community to abolish weapons of mass destruction.

- India's primary objective is to achieve economic, political, social, scientific and technological development within a peaceful and democratic framework. This requires an environment of durable peace and insurance against potential risks to peace and stability. It will be India's endeavour to proceed towards this overall objective in cooperation with the global democratic trends and to play a constructive role in advancing the international system toward a just, peaceful and equitable order.
- Autonomy of decision-making in the developmental process and in strategic matters is an inalienable democratic right of the Indian people. India will strenuously guard this right in a world where nuclear weapons for a select few are sought to be legitimised for an indefinite future, and where there is growing complexity and frequency in the use of force for political purposes.
- India's security is an integral component of its development process. India continuously aims at promoting an ever-expanding area of peace and stability around it so that developmental priorities can be pursued without disruption.
- However, the very existence of the offensive doctrine pertaining to the first use of nuclear weapons and the insistence of some nuclear weapons states on the legitimacy of their use even against non-nuclear weapon countries constitute a threat to peace, stability and sovereignty of states.
- This document outlines the broad principles for the development, deployment and employment of India's nuclear forces. Details of policy and strategy concerning force structures, deployment and employment of

nuclear forces will flow from this framework and will be laid down separately and kept under constant review.

- Objectives
  - In the absence of global nuclear disarmament, India's strategic interests require effective, credible nuclear deterrence and adequate retaliatory capability should deterrence fail. This is consistent with the United Nations (UN) Charter, which sanctions the right of self-defence.
  - The requirements of deterrence should be carefully weighed in the design of Indian nuclear forces and in the strategy to provide for a level of capability consistent with maximum credibility, survivability, effectiveness, safety and security.
  - India shall pursue a doctrine of credible minimum nuclear deterrence. In this policy of 'retaliation only', the survivability of our arsenal is critical. This is a dynamic concept related to the strategic environment, technological imperatives and the needs of national security. The actual size, components, deployment and employment of nuclear forces will be decided in the light of these factors. India's peacetime posture aims at convincing any potential aggressor that (a) any threat of use of nuclear weapons against India shall invoke measures to counter the threat; and (b) any nuclear attack on India and its forces shall result in punitive retaliation with nuclear weapons to inflict damage unacceptable to the aggressor.
  - The fundamental purpose of Indian nuclear weapons is to deter the use and threat of use of nuclear weapons by any state or entity against India and its forces. India will not be the first to initiate a nuclear strike, but will respond with punitive retaliation should deterrence fail.

- India will not resort to the use or threat of use of nuclear weapons against states which do not possess nuclear weapons, or are not aligned with nuclear weapon powers.
- Deterrence requires that India maintain:
    - Sufficient, survivable and operationally prepared nuclear forces.
    - A robust command and control system.
    - Effective intelligence and early warning capabilities.
    - Comprehensive planning and training for operations in line with the strategy.
    - The will to employ nuclear forces and weapons.
- Highly effective conventional military capabilities shall be maintained to raise the threshold of outbreak, both of conventional military conflict as well as that of threat or use of nuclear weapons.

- Nuclear Forces
    - India's nuclear forces will be effective, enduring, diverse, flexible and responsive to the requirements in accordance with the concept of credible minimum deterrence. These forces will be based on a triad of aircraft, mobile land-based missiles and sea-based assets in keeping with the objectives outlined above. Survivability of the forces will be enhanced by a combination of multiple redundant systems, mobility, dispersion and deception.
    - The doctrine envisages assured capability to shift from peacetime deployment to fully employable forces in the shortest possible time, and the ability to retaliate effectively even in a case of significant degradation by hostile strikes.
- Credibility and Survivability: The following principles are central to India's nuclear deterrent:

  - Credibility: Any adversary must know that India can and will retaliate with sufficient nuclear weapons to inflict destruction and punishment that the aggressor will find unacceptable if nuclear weapons are used against India and its forces.
  - Effectiveness: The efficacy of India's nuclear deterrent be maximised through synergy among all elements involving reliability, timeliness, accuracy and weight of the attack.
- Survivability
  - India's nuclear forces and their command and control shall be organised for very high survivability against surprise attacks and for rapid punitive response. They shall be designed and deployed to ensure survival against a first strike and to endure repetitive attrition attempts with adequate retaliatory capabilities for a punishing strike which would be unacceptable to the aggressor.
  - Procedures for the continuity of nuclear command and control shall ensure a continuing capability to effectively employ nuclear weapons.
- Command and Control
  - Nuclear weapons shall be tightly controlled and released for use at the highest political level. The authority to release nuclear weapons for use resides in the person of the Prime Minister of India or the designated successor(s).
  - An effective and survivable command and control system with requisite flexibility and responsiveness shall be in place. An integrated operational plan, or a series of sequential plans, predicated on strategic objectives and a targeting policy shall form part of the system.
  - For effective employment, the unity of command and control of nuclear forces including dual capable delivery systems shall be ensured.

- The survivability of the nuclear arsenal and effective command, control, communications, computing, intelligence and information (C4I2) systems shall be assured.
- The Indian defence forces shall be in a position to execute operations in a Nuclear, Biological and Chemical (NBC) environment with minimal degradation.
- Space-based and other assets shall be created to provide early warning, communications and damage/detonation assessment.

- Security and Safety
  - Security: Extraordinary precautions shall be taken to ensure that nuclear weapons, their manufacture, transportation and storage are fully guarded against possible theft, loss, sabotage, damage or unauthorised access or use.
  - Safety is an absolute requirement and tamper-proof procedures and systems shall be instituted to ensure that unauthorised or inadvertent activation/use of nuclear weapons does not take place and risks of accident are avoided.
  - Disaster Control: India shall develop an appropriate disaster control system capable of handling the unique requirements of potential incidents involving nuclear weapons and materials.
- Research and Development
  - India should step up efforts in research and development to keep up with technological advances in this field.
  - While India is committed to maintain the deployment of a deterrent which is both minimum and credible, it will not accept any restraints on building its R&D capability.

- Disarmament and Arms Control
  - Global, verifiable and non-discriminatory nuclear disarmament is a national security objective. India shall continue its efforts to achieve the goal of a nuclear weapon-free world at an early date.
  - Since no-first use of nuclear weapons is India's basic commitment, every effort shall be made to persuade other states possessing nuclear weapons to join an international treaty banning first use.
  - Having provided unqualified negative security assurances, India shall work for internationally binding unconditional negative security assurances by nuclear weapon states to non-nuclear weapon states.
  - Nuclear arms control measures shall be sought as part of the national security policy to reduce potential threats and to protect our own capability and its effectiveness.
  - In view of the very high destructive potential of nuclear weapons, appropriate nuclear risk reduction and confidence-building measures shall be sought, negotiated and instituted.

APPENDIX B

Opening Remarks by National Security Advisor Brajesh Mishra at the Release of Draft Indian Nuclear Doctrine on 17 August 1999

Ladies and Gentlemen,

I am happy to present to you the draft of the Nuclear Doctrine prepared by the National Security Board. A copy has been placed in each of the seats in the hall. We have decided to make this document public in keeping with our position in favour of greater transparency in decision-making. Please note that this is a draft proposed by the NSAB and has not yet been approved by the government. That will have to wait until after the General Elections.

As our thinking on the nuclear tests has been fairly well publicised, I do not intend to go over the ground again. Suffice it to say that this was a step necessitated by the security environment and our need to ensure for ourselves the element of strategic autonomy in decision-making which we will need in the coming years. Our position has all along been that global security would be enhanced by the universal elimination of all nuclear weapons, and this remains our conviction today. Unfortunately, the indefinite extension of the Non-Proliferation Treaty in 1995 was in the reverse direction.

Our nuclear weapons are not country-specific but, as I mentioned earlier, are aimed at providing us the autonomy of exercising strategic choices in the best interest of our country, without fear or coercion in a nuclearised environment. That being so, we have adopted a policy of minimum deterrence as the basic building block of our nuclear thinking. Minimum but credible deterrence is the watchword of our nuclear doctrine. From this, flows the decision to adopt a no-first-use posture. We have, therefore, given unconditional guarantees to states that do not have nuclear weapons, or are not aligned with nuclear weapon powers.

A cardinal principle regarding the use of nuclear weapons is that of civilian control. Only the elected civilian leader of the country is empowered to authorise the use of nuclear weapons. As the recent operations in Kargil have demonstrated, our system and the political leadership believe with (in) great responsibility and restraint, as you would expect from the largest democracy in the world. This sense of responsibility will also guide our actions with regard to nuclear weapons.

With these words, I have great pleasure in releasing the document for public discussion and debate.

Appendix C

Press Release of CCS Meeting on Nuclear Doctrine

The latest official communication regarding India's nuclear doctrine was through a press release.[1] It stated: 'On 4 January 2003, the Cabinet Committee on Security (CCS) met to review the progress in the operationalising of India's nuclear doctrine.

1. The Committee decided that the following information, regarding the nuclear doctrine and operational arrangements governing India's nuclear assets, should be shared with the public.
2. India's nuclear doctrine can be summarised as follows:

- Building and maintaining a credible minimum deterrent;
- A posture of 'no first use'. Nuclear weapons will only be used in retaliation against a nuclear attack on Indian territory or on Indian forces anywhere;
- Nuclear retaliation to a first strike will be massive and designed to inflict unacceptable damage.
- Nuclear retaliatory attacks can only be authorised by the civilian political leadership through the NCA.
- Non-use of nuclear weapons against non-nuclear weapon states;

---

[1] Ministry of External Affairs, Media Centre, 4 January 2003, 'The Cabinet Committee on Security Reviews Operationalization of India's Nuclear Doctrine'. www.mea.gov.in/press-releases.htm?dtl/20131/

- However, in the event of a major attack against India, or Indian forces anywhere, by biological or chemical weapons, India will retain the option of retaliating with nuclear weapons;
- A continuance of strict control on export of nuclear- and missile-related materials and technologies, participation in the Fissile Material Cut-off Treaty negotiations and continued observance of the moratorium on nuclear tests.
- Continued commitment to the goal of a nuclear weapon-free world, through global, verifiable and non-discriminatory nuclear disarmament.

3. The NCA comprises a political council and an executive council. The Prime Minister chairs the political council. It is the sole body which can authorise the use of nuclear weapons.
4. The executive council is chaired by the national security advisor. It provides inputs for decision-making by the NCA and executes the directives given to it by the political council.
5. The CCS reviewed the existing command and control structures, the state of readiness, the targeting strategy for a retaliatory attack and operating procedures for various stages of alert and launch. The Committee expressed satisfaction with the overall preparedness. The CCS approved the appointment of a Commander-in-Chief, SFC, to manage and administer all strategic forces.
6. The CCS also reviewed and approved the arrangements for alternate chains of command for retaliatory nuclear strikes in all eventualities.'

INDEX